Path Planning for Vehicles Operating in Uncertain 2D Environments

T0296650

Path Planning for Vehicles Operating in Uncertain 2D Environments

Viacheslav Pshikhopov
Southern Federal University, Taganrog, Russia

With Contributions By

Aleksey Pyavchenko
Southern Federal University

Evgeny Kosenko
Southern Federal University

Igor Shapovalov
Southern Federal University

Mikhail Medvedev
Southern Federal University

Roman Saprykin
Southern Federal University

Valery Finaev
Southern Federal University

Victor Krukhmalev
RoboCV, Ltd.

Victor Soloviev
Southern Federal University

Vladimir Pereverzev
Southern Federal University

Viacheslav Guzik
Southern Federal University

Denis Beloglazov
Southern Federal University

Butterworth-Heinemann
An imprint of Elsevier
elsevier.com

Butterworth-Heinemann is an imprint of Elsevier
The Boulevard, Langford Lane, Kidlington, Oxford OX5 1GB, United Kingdom
50 Hampshire Street, 5th Floor, Cambridge, MA 02139, United States

Notices
Knowledge and best practice in this field are constantly changing. As new research
and experience broaden our understanding, changes in research methods,
professional practices, or medical treatment may become necessary.

Practitioners and researchers must always rely on their own experience and
knowledge in evaluating and using any information, methods, compounds, or
experiments described herein. In using such information or methods they should
be mindful of their own safety and the safety of others, including parties for whom
they have a professional responsibility.

To the fullest extent of the law, neither the Publisher nor the authors, contributors,
or editors, assume any liability for any injury and/or damage to persons or
property as a matter of products liability, negligence or otherwise, or from any use
or operation of any methods, products, instructions, or ideas contained in the
material herein.

Library of Congress Cataloging-in-Publication Data
A catalog record for this book is available from the Library of Congress

British Library Cataloguing-in-Publication Data
A catalogue record for this book is available from the British Library

ISBN: 978-0-12-812305-8

For information on all Butterworth-Heinemann publications
visit our website at https://www.elsevier.com/books-and-journals

Working together
to grow libraries in
developing countries

www.elsevier.com • www.bookaid.org

Publisher: Joe Hayton
Acquisition Editor: Sonnini R. Yura
Editorial Project Manager: Intern Gabriela D. Capille
Editorial Project Manager: Mariana Kühl
Production Project Manager: Mohanapriyan Rajendran
Designer: Matthew Limbert

Typeset by TNQ Books and Journals

Contents

List of Contributors

D. Beloglazov
Southern Federal University, Taganrog, Russia

V. Finaev
Southern Federal University, Taganrog, Russia

V. Guzik
Southern Federal University, Taganrog, Russia

E. Kosenko
Southern Federal University, Taganrog, Russia

V. Krukhmalev
RoboCV, Ltd., Moscow, Russia

M. Medvedev
Southern Federal University, Taganrog, Russia

V. Pereverzev
Southern Federal University, Taganrog, Russia

V. Pshikhopov
Southern Federal University, Taganrog, Russia

A. Pyavchenko
Southern Federal University, Taganrog, Russia

R. Saprykin
Southern Federal University, Taganrog, Russia

I. Shapovalov
Southern Federal University, Taganrog, Russia

V. Soloviev
Southern Federal University, Taganrog, Russia

Acknowledgment

This research was supported by the grant of the Russian Foundation for Basic Researches "Development of theory and methods of creation of intelligent position-path control systems for mobile objects under the conditions of environmental uncertainty" (14-19-01533). The research project was conducted at Southern Federal University in Russia.

Abbreviations

UAV	Unmanned aerial vehicles
RAS	Russian Academy of Sciences
GWF	Generalized work functional
SC	Safety criterion
PL	Path length
TCT	Task Completion Time
MSC	Mission success coefficient
DVH-NN	Distance Vector Histogram-Neural network
SAS	Structure of afferent synthesis
DMS	Special decision-making scheme
TVS	Technical vision system
ISPPC	Intelligent system of position path control
NNPP	Neural networking path planner
VFH	Vector field histogram
CDM	Course distances maximum
VMPI	Vehicle motion permission indicator
ADM	Minimal apparent distance
RHI	Right hand indicator
FN	Formal neuron
NUCAC	Neural-like unit for course angle calculation
SS	Space simulation
PE	Processing element
EMS	Environment model subsystem
BNNP	Bionic neural networking planner
EMFS	Environment model formation subsystem
CNN	Convolutional neural networks
GPCU	Global path construction unit
FM	Feature map
OBD	Optimal brain damage
RMSD	Root-mean-square deviation
FPGA	Field-programmable gate array
LV	Linguistic variables
FV	Fuzzy variable
DAMN	Distributed architecture for mobile navigation
CS	Coordinates system
GA	Genetic algorithm
RMSE	Root-mean-square error
EVT	Extended voronoi transform
RRT	Rapidly exploring random tree

Introduction

Vehicles are being used everywhere and the sphere of their application is constantly growing. This makes it essential to design them as autonomous systems. The key characteristics of such vehicles would be high degree of autonomy that is independent of disturbances, ability for a goal search, increased range of action, simplicity, and effective usage.

Increased degree of autonomy calls for necessity to account for a number of features connected to building the mathematical models of a vehicle itself and its environment. The functional capabilities of vehicle control systems should be extended as well.

First, it is necessary to use dynamic vehicle and environment models that should be adequate in the whole functioning range. They must include equations of kinematics and dynamics, actuator equations, and models of environmental interaction and sensor system.

Second, the autonomy increase requires solution of important tasks related to its function in uncertain environments and required adaptation to the changing states of the environment.

Third, autonomous functioning requires development of a vehicle control system using modern methods that would ensure not only actuators and motion control but also autonomous decision making and actions planning.

Therefore, there is a major problem to increase the degree of vehicle autonomy that can be solved by the intellectual technologies used on all the levels of control system that include:

- goal setting level ensuring formation of goal-functions, criteria, estimation of task implementability, interaction with the strategic control level, and goal correction;
- level of control, navigation, and communication of a single vehicle; and
- actuator level.

This book interprets the intelligent technologies that allow realization of human behavioral functions, such as adaptation to uncertain environment, ability to estimate and model the current state, perform goal-setting functions, and action planning.

A vehicle moving in an uncertain environment with obstacles needs a control system that would exactly implement complicated curves. That is why a lower-level control of a vehicle imposes heavy demands on the class of the tasks to be solved. Here we propose to implement a lower level of the control system in a class of position-path systems that allow following the paths that are set by linear and quadratic forms of external vehicle coordinates. This class of paths allows ensuring smooth and precise vehicle motion in an obstructed environment. In addition, linear and quadratic forms give us a convenient tool for formal cohering of the planning level and lower actuating level.

It should be mentioned that the controller level of the control system can be implemented not only using position-path control but also by other methods. However, if the control system uses path representation by a set of points, the controller should include an interpolator. Also, if the controller level requires motion planning in internal coordinates, the controller should have an additional unit for solving inverse kinematics problem.

Implementation of lower levels of vehicle control systems in a class of intelligent technologies and design of all the control levels in a single intelligent basis will be considered in the future works. This book addresses the planning level forming the vehicle trajectories on a surface with stationary obstacles.

In the first chapter we cover the fundamentals of position-path control, present the tasks to be solved, define the control plant, functioning environment, test scenes, and the general structure of the control system. The main definitions are given. The performance criteria are introduced to estimate all the considered path planning methods.

The second chapter considers vehicle path planning methods based on neural networks and neural-like structures. Two methods were tested on the example scenes. First method is an implementation of a planner in a classical neural network based on a class of algorithms without memory. Then the method was modified to be implemented by neural-like structures using formal neurons with step functions of activation.

The third chapter considers the questions of applying the formalism of fuzzy logic for solution of the vehicle path planning tasks. Based on this formalism a method of vehicle's surface motion planning is proposed. Then this method was applied to test scenes.

The fourth chapter presents the developed vehicle's motion path planning method using genetic algorithms for searching on a 2D graph. Two method variants are presented—one with mapping and another without.

The fifth chapter covers the research of graph-analytic trajectory planning methods based on potential fields and Voronoi diagrams. Six widespread algorithms based on potential-field method are presented. There are also two variants of Voronoi diagrams application for vehicle path planning—with mapping and without. The chapter also presents a survey of methods accounting limitations imposed by the vehicle dynamics.

The sixth chapter presents development of a vehicle path planning method basing on a bionic approach using unstable modes for obstacle avoidance. A hybrid algorithm is proposed that uses unstable modes and a concept of a virtual goal point.

The summary presents a comparison of methods considered in this book using integrated performance criteria.

Chapter One

Position-Path Control of a Vehicle

V. Pshikhopov, M. Medvedev

Southern Federal University, Taganrog, Russia

1.1 MOTION-CONTROL SYSTEMS PROBLEMS ANALYSIS

Motion control is one of the most developing areas stimulating improvement of control theory and its applications. Every new generation of vehicles introduces new requirements to conditions, modes, and quality of their functioning. The range of tasks becomes wider and wider. The diversity of vehicles, their environments, and the tasks brings out a wide spectrum of practical and theoretical problems emerging in motion-control systems development.

For example, the work [1] mentions the necessity of developing motion-control automation because of unfavorable human influence in critical situations under short time and with large amounts of information. It also indicates that the creation of perspective vehicles requires the increase of accuracy and speed, invariance to disturbances, and obtaining the qualities of adaptivity and autonomy. These should be achieved under the conditions of incompleteness of a priori information and in uncertainty of external disturbances and environment.

As a result, providing the autonomy to vehicles becomes the main task. This implies a powerful system of path planning and motion performing [2].

An analogous analysis was implemented in Ref. [3]. It is mentioned that under the conditions of increasing requirements to vehicle's functional capabilities, informatization is not sufficient. The center point is the development of planning and control algorithms ensuring a high level of autonomy and adaptivity; algorithms that should be highly effective under the conditions of information incompleteness and uncertainty of environment and external disturbances. Urgent necessity for creation of new vehicle control systems is confirmed by the example of creating specific robotic devices. In the United States, creation of various robotic systems is the center point to the development of the US military forces. The work [3] shows the development examples of a tactical unmanned aerial vehicle (UAV) and an unmanned stratospheric lighter-than-air platform.

A review of UAV was done based on online publications by Sokolov and Teryaev [4]. This report demonstrates the importance of UAV creation for

Path Planning for Vehicles Operating in Uncertain 2D Environments. http://dx.doi.org/10.1016/B978-0-12-812305-8.00001-6

various purposes mentioning that the piloted aviation would be replaced by unmanned one in perspective. The authors of Ref. [4] highlighted the four stages of a single UAV development and also considered the creation of a complex of independent UAVs, interacting UAV groups, and UAVs embedded into complicated functional systems. UAVs can be used for the solution of a wide circle of tasks that include surveillance, monitoring, and reconnaissance; establishing communication, road traffic control, and objects state monitoring, etc. At present not only unmanned planes but the helicopters are also being created. Long distance UAV control systems are being developed. The review Ref. [4] leads to conclusion that the UAVs created today need to be distantly controlled and creation of an autonomous UAV is an urgent task.

The most universal solution of the motion-control task is required for the mobile robots. The XII All-Russian meeting on control problems considered the important directions in mobile robots control systems development [5]. In the work [6], we find the following main directions for development in this field:

- control in aviation and astronautics;
- marine vehicles control;
- mechatronics, control, and information processing in robotics; and
- vehicles navigation.

The plenary reports indicated urgency of the tasks related to vehicle navigation, multi-goal control of vehicles functioning in different modes, and organization of flight control for modern space vehicles. The themes of section reports make it clear that the following problems are still important: high-precision maneuvers control; broadening the functioning modes range by application of more detailed mathematical models; increasing autonomy of the existing vehicles; and giving the control systems intellectual qualities and cohering various levels of these systems.

The mentioned tasks were considered in the XII All-Russian meeting for a wide spectrum of vehicles including marine vehicles, space vehicles, airfoil boats, quadcopters, UAVs, and others.

A special attention was paid to the functioning of vehicles in uncertain conflict environments such as the ones with moving and stationary obstacles. Importance and topicality of this problem for autonomous vehicles was also mentioned.

The section "Mechatronics, control and information processing in robotic systems" of XII All-Russian meeting on control problems pointed out that the main solution in important vehicle planning and control problems is to use intellectual technologies including fuzzy logic and neural networks. In Ref. [6] it is mentioned that "…at the present time around the world… an autonomous vehicle control theory is being actively developed together with the theory and methods of information processing in navigation systems under the conditions of uncertainty and noises being present."

In order to stress its importance for autonomous vehicles, a separate direction called "Intellectual systems in control" was created. Inside this direction, the following questions were considered: multiagent systems; methods of neural networks tuning; control of robots and their groups; application of fuzzy logic and cognitive maps. The task of vehicle control, and particularly flying vehicle control, intellectualization was called one of the most urgent ones. The second in the list was autonomous robots group control.

A lot of attention was paid to the tasks of vehicles trajectory planning and control at the 19th world IFAC congress held in Cape Town (SAR) in August 2014. Just the plenary session on robot control and intellectual system had three reports (with a total of 8). More than 25 sections of the congress were devoted to mobile robots and vehicles control, planning, navigation, and intellectual control methods for vehicles. Besides, a large number of reports were devoted to specific problems addressing the features of vehicle's environment. The congress report topics emphasized the high importance of vehicle planning and control problems.

In work [7] there is a brief survey of achievements of V.A. Trapeznikov Institute of Control Sciences (RAS) in the area of reference model adaptive control systems and space vehicles control. The scientists from this institute proposed to use a reference model adaptation for building rocket control systems. The first few works on this topic were published in 1965 [8,9]. A series of fundamental results were achieved using the direct Lyapunov method for adaptation algorithms synthesis [10]. The authors of the survey [7] highlighted the problems related to ensuring asymptotic stability of the synthesized adaptive systems and to the possibility of parameters identification. Based on the adaptation ideal, the ICS team proposed a concept of coordinate-parametric control [11] in which flying vehicle control is performed both by the traditional control devices and by changing the aircraft's configuration, e.g., changing air rudders area, changing distance between centers of mass and pressure, etc. Later they introduced a principle of adjustable operability based on the solution of multicriterial synthesis task accounting for a large number of performance criteria such as stability, coordinates limitations, invariance, autonomy, optimality by some criterion, etc. The mentioned criteria form a complex set of requirements put on a modern control system.

In works [12,13] there is a review of intelligent vehicle control systems. Based on information acquired from the electronic sources, the authors perform an analysis of global developments in robotics. The performed analysis allowed finding a number of tendencies that include: revitalization of works on creation of robots for air, ground, and underwater deployment; reducing the sizes of a number of robotic devices and usage of artificial intelligence technologies in creation of autonomous objects. These technologies are used for setting and correcting the control goals and programs for realization of these goals. They are also used for creation of control algorithms under the uncertainty conditions caused by various factors in actuators, motion control, and behavior planning subsystems.

The importance of giving intellectual qualities to the vehicle control systems is also mentioned in Ref. [14]. The authors of this work indicated the necessity to use intelligent technologies for solution of vehicle control tasks under the conditions of environmental uncertainty, high speed of controlled processes, extreme conditions, and active counteraction. The area of intelligent systems application is defined for conditions of a priori uncertain environment and for a significantly varying situation. In Ref. [14] it is intended to give intelligent qualities not only at the stage of behavior planning but also at the tactical control level for solution of motion-control task for a dynamic object in an obstructed and changing environment.

Let us note that attempts to apply artificial intelligence technologies for solution of traditional control tasks at a lower level are encountering the limitations of the modern motion-control systems. However, despite the significant increase of publications on intelligent systems, there are very few examples of their application for control of complicated dynamic objects [15].

The work [16] makes a review of various approaches to non-adaptive robots motion-control methods in a clear environment. A task of motion along a set trajectory is considered for a case when it is set in a stationary coordinate system with respect to the goal. In another case, the goal location cannot be known in advance and is determined by the sensor system. In Ref. [16] two types of motion are considered—motion to a preset position and path-following. Motion-control methods are separated into dynamic and non-dynamic. In case of non-dynamic control laws, the results are satisfactory for positioning control tasks and for following slow paths. For exact following of fast trajectories, it is necessary to use dynamic control. The work [17] also mentions the problem in setting a path. It is indicated that the best way is to control the robot in external coordinates and such systems are still at the research stage. The authors of Ref. [16] highlight five main directions in non-adaptive robot control methods: (1) optimal control; (2) inverse problems; (3) control decomposition; (4) force feedback; and (5) decentralized control. For the optimal control methods, the work notes a number of problems related to the usage of a linearized model and low robustness. For inverse problems and decentralized control, there are problems related to computing the full dynamic model in real time, robustness of the obtained control laws, and to selection of controller coefficients. For the control decomposition approach, it is necessary to perform trajectory recalculation for the joints coordinates and this approach has a series of disadvantages including high sensitivity to changes of parameters. Using force feedbacks for control of manipulator's motion, allows compensating the robot's dynamics but requires presence of hinge moment sensors leading to a number of technological problems related to the reduction of manipulator links stiffness.

Academician A.A. Krasovsky has developed an approach to control synthesis based on generalized work functional (GWF) [18,19]. Many variants of optimal adaptive flying vehicle control were developed using GWF. It should be mentioned that this approach possesses the general drawback of the quadratic criteria application for multiply connected nonlinear systems with undefined parameters that lead to equations that cannot be solved analytically. However, for some cases there is an analytical solution, e.g., the one obtained in Ref. [20].

In the work [21] it is noted that current remote control systems have a number of weaknesses limiting the area of their applications. The authors indicated the necessity to increase the degree of autonomy of the mobile ground platforms and highlight a series of unsolved tasks including planning, and following trajectories on operational and tactical levels.

Thus analysis shows that at present, the urgent tasks in motion control include building control for multiply connected nonlinear systems autonomously functioning under the conditions of uncertain parameters, and disturbances and presence of stationary or mobile obstacles. For the proper functioning of autonomous objects, the crucial importance lies at the motion planning level that should ensure formation of intermediate goals, and functioning of the vehicle in the environments with stationary and mobile goals.

The structure of an autonomous vehicle's motion-control system should include an intelligent planning level in coherence with a controller lever.

In this monograph, such a structure is implemented in a system of position-path control that allows implementation of the controller level and a number of functions at the planning level. Moreover, the position-path control system allows for effective controller—planner coupling.

Synthesis of systems of intelligent planning and position-path motion control is performed consecutively and includes the following stages: building and analysis of vehicle's mathematical model, synthesis of motion planning and control algorithms, analysis of the closed-loop control system, and technical implementation.

1.2 MATHEMATICAL MODELS OF MOTION

Solution of control problems, not only for vehicles but also for the other types of objects, includes certain stages. And the first one, that to a large extent determines the statement of control problem, is the derivation of correct mathematical models describing a certain real object with a various degree of adequacy.

A variety of works are devoted to derivation and analysis of vehicles mathematical models. They considered various aspects of model-building procedures in relation to vehicles [22—30].

1.2.1 Vehicle's Mathematical Model

The tasks solved by the modern vehicle's control systems require functioning in the modes that reveal qualities of multiple connectivity and nonlinearity of the object. This section considers the procedure of mathematical model deriving for a vehicle moving on a surface. This is based on a solid body model, which describes multiply connected nonlinear motion with a high degree of adequacy.

The vehicle's mathematical model synthesis procedure includes:
- derivation of the vehicle's kinematic model determining its position and orientation in a selected reference system; and
- derivation of the dynamics model determining external and internal forces and torques influencing the vehicle during its motion.

The successful solution of the vehicle's mathematical model synthesis task significantly influences the qualitative characteristics of the control system.

The main elements of the vehicle's space motion mathematical model are:

- kinematic equations of forward and rotary motion in a normal terrestrial reference system;
- dynamic equations in projections on the axes of the vehicle-fixed coordinate system with its origin located at the vehicle's center of gravity; and
- expressions for calculation of projections of external forces and torques.

Mathematical model of a vehicle in a matrix form can be presented in the following equation [28,29,31]:

$$
\begin{aligned}
\dot{y} &= R(\Theta)x, \\
\dot{x} &= M^{-1}(F_u(\delta) - F_d - F_v), \\
\dot{\delta} &= -T_\delta \delta + K_\delta u,
\end{aligned}
\tag{1.1}
$$

where $y = [P\Theta]^T$ is the vector of linear and angular vehicle positions in terrestrial coordinate system; x, vector of linear and angular velocities in a vehicle-fixed coordinate system; δ, vector of vehicle actuators' state coordinates; $R(\Theta)$, kinematics matrix depending on vector of angular coordinates Θ; M, matrix of vehicle's inertial parameters; $F_u(\delta)$, vector of controlling forces and torques; F_d, vector of dynamic forces and torques acting on a vehicle; F_v, vector of external disturbances; T_δ, inertia matrix of actuators; K_δ, transfer coefficients matrix of actuators; u, control vector.

The set of coordinates that uniquely describes the vehicle's position in the space of external coordinates will be called a vector of external coordinates y. It is assumed that all its elements are measurable. Whereas a set of controlled coordinates (driving wheels rotation speed, driven wheel angle, robot's angular velocities, etc.) will be called a vector of internal coordinates and will also be assumed to have all measurable components.

According to Eq. (1.1) the vehicle's mathematical model structure is presented in Fig. 1.1.

It should be mentioned that generally models of vehicle's dynamics in the form of Eq. (1.1) are multiply connected systems of nonlinear differential equations with their elements defined by composition and parameters of a specific vehicle, and by the structure and character of external disturbances.

The model based on Eq. (1.1) is the basic model for building the vehicle's motion-control system. Usually this system is simplified. For example, a common technique of motion separation [32] can be used. It allows decomposition of the control system into a set of local controllers. If dynamics of the actuators is much faster than dynamics of the vehicle itself, the control system is built using a servomechanism principle, which is acceptable in most cases. For a vehicle moving at a low speed it is possible to use only kinematic relations. This simplifies the control task solution but limits the control system's range of operation. Another essential and frequently used simplification is the linearization of the motion model [33−37]. This simplification is allowed in the

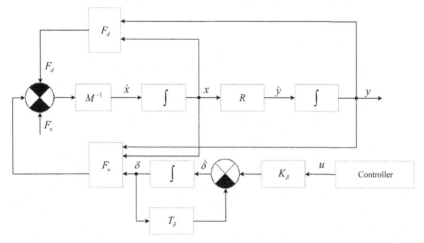

FIGURE 1.1 Vehicle's mathematical model general structure.

region of small deviations from the desired functioning mode. So the autopilot should have various sets of linearized models for each functioning mode of a vehicle.

In vehicle's control system synthesis, it is frequently assumed that there is an analytical or numeric solution of its kinematic equations. In this case the control system is build using only dynamic equations [38,39]. Such an approach leads to reduction of system's accuracy under the action of disturbances.

1.2.2 Mathematical Model of a Wheeled Vehicle

The general structure of a vehicle's equations in a two-dimensional space matches that of Eq. (1.1), i.e., it consists of motion dynamics equations, kinematics equations, and drives equations.

The type of chassis' kinematic schema of a mobile robot determines its maneuverability and to a large extent its controllability. This introduces certain specific features both into path planning and control algorithms synthesis procedures.

Mathematical models of chassis of wheeled vehicles, mobile robots, and automobiles are considered in detail in works [22,23,40]. Usually a wheeled robot's model is built based up on the solid body model or a system of solid bodies connected by joints [22,23,40]. One solid body is the frame, and the rest are the attached wheels. The frame's position is defined in a certain basic coordinate system and the wheel positions by the rotation angle with respect to the axis fixed to the frame.

In this monograph, a controlled vehicle is represented by a tricycle controlled by changing the rotation velocities of the wheels located on one axle, while the third wheel is passive (tank type). The work [22] gives the kinematic scheme of a chassis presented in Fig. 1.2. The front wheel is not the controlling one.

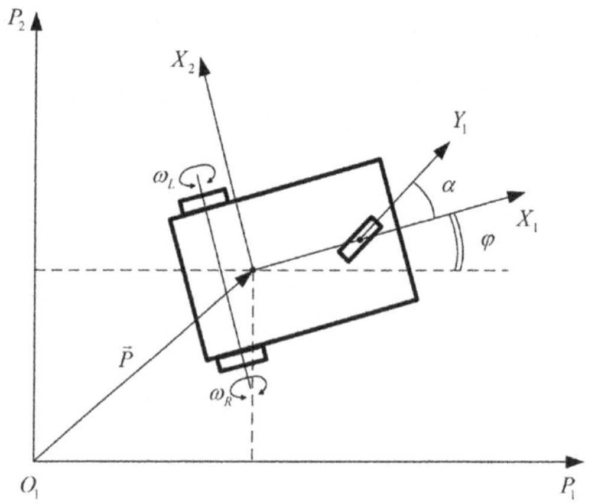

FIGURE 1.2 Tank-type vehicle.

In Fig. 1.2 $\{O_1P_1P_2\}$ is a fixed system of coordinates for determining the frame's position.

Nonholonomic links put on the system result in the following kinematic model written with respect to wheels angular velocities:

$$\dot{y} = R(\Theta)x,$$
$$\dot{x} = B_u\delta - F_d, \tag{1.2}$$

$$y = \begin{bmatrix} P_1 \\ P_2 \end{bmatrix}, \quad x = \begin{bmatrix} \omega_L \\ \omega_R \end{bmatrix}, \quad \delta = \begin{bmatrix} u_L \\ u_R \end{bmatrix},$$

$$R = \begin{bmatrix} \cos(\varphi) & -\sin(\varphi) \\ \sin(\varphi) & \cos(\varphi) \end{bmatrix} \begin{bmatrix} 0,5r & 0,5r \\ -\dfrac{ar}{2b} & \dfrac{ar}{2b} \end{bmatrix} \tag{1.3}$$

$$B_u = \left(\begin{bmatrix} 0,5mr & 0,5mr \\ -\dfrac{Jr}{2b} & \dfrac{Jr}{2b} \end{bmatrix} \right)^{-1} \begin{bmatrix} -\dfrac{1}{r} & -\dfrac{1}{r} \\ \dfrac{b}{r} & -\dfrac{b}{r} \end{bmatrix} \begin{bmatrix} d_{11}^1 & 0 \\ 0 & d_{22}^1 \end{bmatrix} \tag{1.4}$$

$$F_d = \dfrac{2}{r^2} \left(\begin{bmatrix} m & m \\ -\dfrac{J}{b} & \dfrac{J}{b} \end{bmatrix} \right)^{-1} \begin{bmatrix} -1 & -1 \\ b & -b \end{bmatrix} \begin{bmatrix} d_{11}^2 & 0 \\ 0 & d_{22}^2 \end{bmatrix} \begin{bmatrix} \omega_L \\ \omega_R \end{bmatrix} \tag{1.5}$$

where angle φ is the orientation angle of the basis $\{O_2X_1X_2\}$ in the coordinate system $\{O_1P_1P_2\}$; r, the wheel radius; a and b, the chassis kinematic parameters; m and J, the reduced chassis mass and moment of inertia; d_{ij}^k, the drive's coefficients; u_L, the controlling voltages of the drives; M_L and M_R, the torques developed by rotors of the left and right drives, respectively, $l = \{L,R\}$.

Eqs. (1.2)−(1.5) present the robot's model assuming that there is no wheel slip and the motion surface is ideally flat.

1.2.3 Description of Vehicle Functioning Test Scenes

In this monograph, the following assumptions are made for the test scenes:
- test scene is a flat surface;
- there is no wheel slip;
- there are obstacles on the surface that can be pointed or have a complicated form. Complicated obstacles are the arrays of pointed obstacles; and
- the moment when the functioning starts, the vehicle has no information about the obstacles location.

In a general case the vehicle's functioning scene can include stationary and mobile obstacles. In all motion planning research work in this book, the obstacles are assumed to be quasi-stationary.

Complicated scenes are considered in this book. The term "complicated scene" means that it can include complicated obstacle-objects that are greater than vehicle's size and have a specific form. In the test scenes, the complicated obstacles are described by uneven and generally non-convex contours consisting of unity-size points that can form line segments.

In model experiments, we used a flat-structured environment 25×25 m in size with several types of obstacles presented in Fig. 1.3 (convex, closely located, L-shaped, and U-shaped). The vehicle moves from initial point $A_0(p_{10}, p_{20})$ to the final one $A_f(p_{1f}, p_{2f})$.

1.3 MOTION PATH PLANNING

In autonomous vehicles control one of the critical tasks is planning the motion path. It is necessary to describe the generated paths by a limited amount of information [41]. In control systems with an operator in control contour (pilot, driver), there is no such problem. Operator generates the path and performs monitoring of the path-following process. However, if the planning and path-following functions are structurally separated, the path description should be cohered with the lower controller level.

Usage of large arrays approximating the path with points is a possible solution for automatic description of the generated path. However, this approach is redundant. In work [42], it is proposed to plan paths in the form of connected straight lines which essentially lowers the class of planned paths and, therefore, vehicle's functional capabilities. In works [43,39] the paths are represented by a sequence of straight lines and arcs of a circle. However, the path's flexion is not continuous and hence, strictly speaking, the vehicle can't move evenly along it.

The authors of the works Refs. [39,44−46] proposed control systems assuming paths planning in the internal coordinates space. Since the transfer procedure from external to internal coordinate's space is hardly formalizable, there will be difficulties in implementation of the desired paths in case of complicated kinematic schemes. The mentioned methods can be used in analysis of the vehicle's motion for the set control actions.

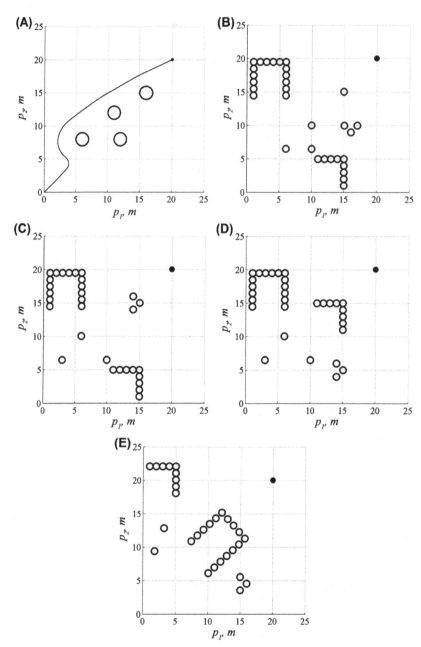

FIGURE 1.3 2D structured environment with obstacles.

The specificity of the vehicle's motion assumes solution of the following tasks in forming the requirements for the steady state modes:

- positioning control task—stabilization at the set point in the space of the basic coordinates A_f and, if necessary, with desired values of orientation angles;

- path control task—motion along the paths set in the space of basic coordinates with a constant speed V and a set orientation of the axes of the vehicle-fixed coordinate system;
- position-path control task—motion to the set point in the space of basic coordinates along the preset path with a set orientation and without additional requirements of speed.

In position-path approach the desired requirements to the path in the space R^n of the working coordinates are set in quadratic and linear forms [28]. During the motion along the surface, they can be presented as follows:

$$\Psi_{tr} = \begin{bmatrix} A_1(t)\mathrm{diag}(P)P + A_2(t)P + A_3(t) \\ \Phi(P, \varphi, t) \end{bmatrix} = 0 \qquad (1.6)$$

$$A_1(t) = \begin{bmatrix} a_{111}(t) & a_{112}(t) \\ a_{121}(t) & a_{122}(t) \end{bmatrix}, \quad A_2(t) = \begin{bmatrix} a_{211}(t) & a_{212}(t) \\ a_{221}(t) & a_{222}(t) \end{bmatrix},$$

$$A_3(t) = \begin{bmatrix} a_{31} \\ a_{32} \end{bmatrix}, \quad P = \begin{bmatrix} p_1 \\ p_2 \end{bmatrix}, \quad \mathrm{diag}(P) = \begin{bmatrix} p_1 & 0 \\ 0 & p_2 \end{bmatrix}$$

where Φ is the function setting the requirements to the vehicle's orientation angle φ; and p_1, p_2, the vehicle's position in the external system of coordinates.

The requirements to motion speed are set in the following form:

$$\Psi_{sp} = J_s \dot{y} + J_t + \tilde{V} = 0, \qquad (1.7)$$

$$J_s = \begin{bmatrix} 2P^T A_1(t) + A_2(t) & 0 \\ \dfrac{\partial \Phi}{\partial P^T} & \dfrac{\partial \Phi}{\partial \Theta^T} \end{bmatrix},$$

$$J_t = \begin{bmatrix} (P^T \dot{A}_1(t) + \dot{A}_{i2}(t))P + \dot{A}_3(t) \\ \dfrac{\partial \Phi_t(P, \Theta, t)}{\partial t} \end{bmatrix},$$

$$\tilde{V} = \begin{bmatrix} 0 & \xi(V^2 - V^{*2}) & 0 \end{bmatrix}^T,$$

where V, V^*, are the vehicle's speed and its desired value, respectively; $\dot{A}_1, \dot{A}_2, \dot{A}_3$, matrices of time derivatives of the matrix elements A_1, A_2, A_3, respectively, or estimates of the changing speeds; and Φ_t, the component of Φ explicitly depending on parameter t.

Obviously the path-following speed should satisfy the following inequality

$$V^* \leq V_{\max} \qquad (1.8)$$

where V_{max} is the maximal path speed determined by power capabilities of the vehicle's actuators.

Depending on a specific task, the manifolds in Eqs. (1.6) and (1.7) are set as follows:

- motion along the straight line $p_2 = kp_1 + b$ going through the two fixed points (p_{10}, p_{20}) and (p_{1f}, p_{2f}) with a constant speed V^*:

$$A_2 = (-k \quad 1), \quad A_3 = b, \quad A_1 = 0,$$

$$\tilde{V} = \left(0 \quad (V^2 - V^{*2})\right)^T, \quad J_s = \begin{vmatrix} -k & 1 & 0 \\ 0 & 0 & 0 \end{vmatrix}, \quad J_t \equiv 0,$$

where $k = \frac{p_{2f} - p_{20}}{p_{1f} - p_{10}}$ and $b = p_{20} - p_{10}k$

- rotation by an angle of α^* during a motion with a motion speed equal to V^*:

$$\Phi_1 = \alpha - \alpha^*, \quad \tilde{V} = \left((V^2 - V^{*2})0\right)^T, \quad J_s = \begin{vmatrix} 0 & 0 & 0 \\ 0 & 0 & 1 \end{vmatrix}, \quad J_t \equiv 0,$$

- transition to a fixed point (p_{1f}, p_{2f}) with a zero final speed:

$$A_1 = \begin{bmatrix} 0 & 0 \\ 0 & 0 \end{bmatrix}, \quad A_2 = \begin{bmatrix} 1 & 0 \\ 0 & 1 \end{bmatrix}, \quad A_3 = \begin{bmatrix} -p_{1f} \\ -p_{2f} \end{bmatrix},$$

$$\tilde{V} \equiv 0, \quad J_s = \begin{vmatrix} 1 & 0 & 0 \\ 0 & 1 & 0 \end{vmatrix}, \quad J_t \equiv 0$$

- following a path going through two nonstationary points, with a constant speed V^*,

$$A_2 = (-k(t) \quad 1), \quad A_3 = b(t), \quad A_1 = 0,$$

$$\tilde{V} = \left(0 \quad (V^2 - V^{*2})\right)^T, \quad J_s = \begin{vmatrix} -k(t) & 1 & 0 \\ 0 & 0 & 0 \end{vmatrix},$$

$$J_t = \begin{vmatrix} -\dot{k}(t)p_1 + p_2 - \dot{b}(t) \\ 0 \end{vmatrix}$$

1.4 ALGORITHMS OF POSITION-PATH CONTROL

This method generally controls the vehicle's motion set by Expressions (1.6) and (1.7), ensuring asymptotic stability of the planned paths [47–51]. In

Ref. [28] the tasks of vehicle's control are solved using its full model Eq. (1.1), by either using dynamics and kinematics equations, or kinematics equations alone.

In this chapter, we present the solution of positioning and path-following tasks for a vehicle described by Eqs. (1.2)–(1.5).

According to Ref. [28], the requirements to vehicle in solution of positioning task are determined by the following expressions:

$$\Psi_{tr} = [A_1(t)\text{diag}(P)P + A_2(t)P + A_3(t)] = 0, \tag{1.9}$$

$$A_1 = \begin{bmatrix} 0 & 0 \\ 0 & 0 \end{bmatrix}, \ A_2 = \begin{bmatrix} 1 & 0 \\ 0 & 1 \end{bmatrix}, \ A_3 = \begin{bmatrix} -p_1^* \\ -p_2^* \end{bmatrix}. \tag{1.10}$$

Let us define the requirements for the closed-loop control system in the form of following reference equation:

$$\ddot{\Psi}_{tr} + T_1\dot{\Psi}_{tr} + T_2\Psi_{tr} = 0, \tag{1.11}$$

where T_1, T_2 are the controller parameters of the matrices.

From Eqs. (1.1)–(1.5), (1.9), and (1.10) and from Eq. (1.11), we get

$$\begin{bmatrix} u_L \\ u_R \end{bmatrix} = -(A_2RB_u)^{-1}\left(A_2\dot{R}\begin{bmatrix} \omega_L \\ \omega_R \end{bmatrix} - A_2RF_d - T_1A_2R\begin{bmatrix} \omega_L \\ \omega_R \end{bmatrix} - T_2\Psi_{tr}\right) \tag{1.12}$$

If control is synthesized using only kinematics equations, the control actions include the wheel rotation velocities and control, and Eq. (1.12) is transformed to the following Eq. (1.13):

$$\begin{bmatrix} \omega_l \\ \omega_r \end{bmatrix} = -\left(-\begin{bmatrix} \cos\varphi & -\sin\varphi \\ \sin\varphi & \cos\varphi \end{bmatrix}\begin{bmatrix} 0.5r & 0.5r \\ \dfrac{0.5ra}{b} & \dfrac{0.5ra}{b} \end{bmatrix}\right)^{-1}\begin{bmatrix} T_1 & 0 \\ 0 & T_1 \end{bmatrix}\left(\begin{bmatrix} p_1 \\ p_2 \end{bmatrix} + A_3\right) \tag{1.13}$$

If the vehicle (Eqs. 1.2–1.5) follows a straight path, the path manifold Eq. (1.6) also has the form of Eq. (1.9). And the matrices Eq. (1.10) take the following form:

$$A_2 = \begin{bmatrix} 1 & -k \end{bmatrix}, \ A_3 = \begin{bmatrix} -b \end{bmatrix}, \tag{1.14}$$

where k, b are path parameters.

If motion is performed with a constant speed, the speed manifold Eq. (1.7) takes the following form:

$$\Psi_{sp} = A_4 \begin{bmatrix} \omega_L \\ \omega_R \end{bmatrix} + A_5 = 0, \quad A_4 = \begin{bmatrix} r & r \end{bmatrix}, \quad A_5 = -V^* \quad (1.15)$$

The requirements for the behavior of the closed-loop system are determined by the reference Eqs. (1.11) and (1.16):

$$\dot{\Psi}_{sp} + T_3 \Psi_{sp} = 0 \quad (1.16)$$

Solving Eqs. (1.11) and (1.16) yields the following equation,

$$\begin{bmatrix} u_L \\ u_R \end{bmatrix} = - \begin{bmatrix} A_2 RB_u & 0 \\ 0 & A_4 B_u \end{bmatrix}^{-1} \left(A_2 \dot{R} \begin{bmatrix} \omega_L \\ \omega_R \end{bmatrix} - A_2 R F_d - T_1 A_2 R \begin{bmatrix} \omega_L \\ \omega_R \end{bmatrix} - T_2 \Psi_{tr} \\ -A_4 B_2 - T_3 \Psi_{sp} \right)$$

$$(1.17)$$

Let us consider an example of position-path control of a vehicle described by Eqs. (1.2)–(1.5). Assume that at the first motion section, it is necessary to perform solution of a path-following task; at the second, position-path task; and at the third, positioning task with the following values of matrices A_{ij} and parameters ξ, V^*, T_i:

1. section 1, motion along an arc of a circle with a set speed

$$\bar{\xi} = 0, \quad \xi = 1, \quad A_{11} = \begin{vmatrix} 1 & 0 \\ 0 & 1 \end{vmatrix}, \quad A_{12} = \begin{vmatrix} -2p_{10} & -2p_{20} \end{vmatrix}, \quad A_{13} = p_{10}^2 + p_{20}^2 - R_0^2,$$
$$p_{10} = 20, \quad p_{20} = 0, \quad R_0 = 20, \quad V^* = 3, \quad s_1 = 1, \quad s_2 = 2$$

2. section 2, motion to the point (40, 40) along a straight line

$$\bar{\xi} = 1, \quad \xi = 0, \quad A_{11} = \begin{vmatrix} 0 & 0 \\ 0 & 0 \end{vmatrix}, \quad A_{12} = \begin{vmatrix} -22 & 11.3 \end{vmatrix}, \quad A_{13} = 431,$$
$$A_{21} = \begin{vmatrix} 0 & 0 \\ 0 & 0 \end{vmatrix}, A_{22} = \begin{vmatrix} 1 & 0 \end{vmatrix}, \quad A_{23} = -40, \quad s_1 = 2.5, \quad s_2 = 4$$

3. positioning to the point (70, 50)

$$\bar{\xi} = 1, \quad \xi = 0, \quad A_{11} = \begin{vmatrix} 0 & 0 \\ 0 & 0 \end{vmatrix}, \quad A_{12} = \begin{vmatrix} 11 \end{vmatrix}, \quad A_{13} = -P^*, \quad P^*$$
$$= (70 \quad 50)^T, \quad s_1 = 3, \quad s_2 = 3$$

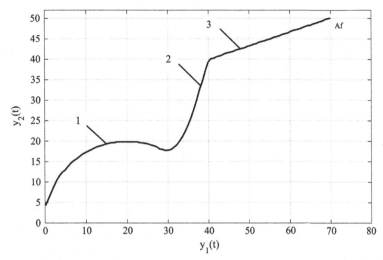

FIGURE 1.4 Vehicle's motion trajectory.

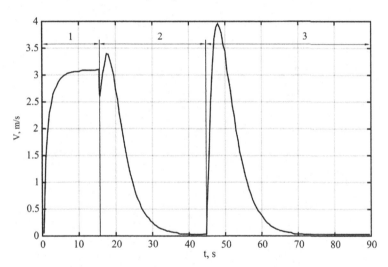

FIGURE 1.5 Vehicle's path-following speed.

The vehicle's motion modeling results are presented in Figs. 1.4—1.10 which fully confirm the possibility to combine various motion modes inside the generalized algorithms in Eqs. (1.12) and (1.17).

1.5 REQUIREMENTS OF PATH PLANNERS

The structure of an intelligent system of position-path control of a vehicle is shown in Fig. 1.11. The vehicle is equipped with a navigation system that can determine its speed and position in external coordinates. The technical vision system, e.g., locator, allows determining the obstacle coordinates in its functioning area. Based on the data from the navigation system and

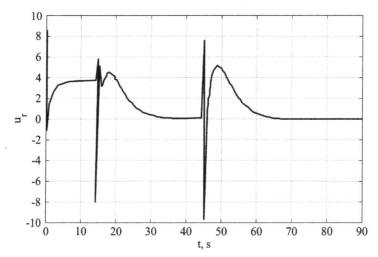

FIGURE 1.6 Right-drive controlling voltage.

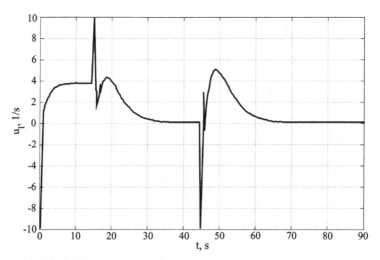

FIGURE 1.7 Left-drive controlling voltage.

locator, the intellectual planner gives the coefficients of quadratic forms to the path-following controller. Generally, the position-path controller gets the sensor data from the actuators.

According to Fig. 1.11 the planner as a part of position-path control system generates the following data:

- matrices of coefficients A_{ij}, and their first and second derivatives in case of mobile obstacles;
- functions Φ reflecting the requirements to the vehicle orientation angles;
- parameter ξ taking the value of "0" (positioning task) or "1" (motion with a set speed); and
- desired value V^* of the vehicle's speed; elements of matrices $\widetilde{C}$, $\widetilde{T}$, $\widetilde{A}$.

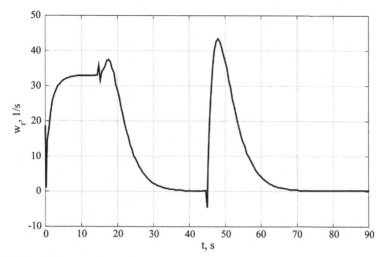

FIGURE 1.8 Right-wheel rotation speed.

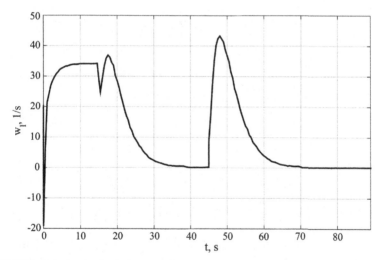

FIGURE 1.9 Left-wheel rotation speed.

In this book, the quality estimation of the planner is performed by the example of a ground-based robot presented in Fig. 1.2, and its kinematic model.

Since only kinematic equations are used for the control system synthesis, the requirements for the vehicle trajectory are formulated as manifolds of Eqs. (1.9) and (1.10). All the planners use the same position-path control law Eq. (1.13).

In general, instead of using the control in Eq. (1.13), any other control law can be used. It is possible to select a control law that performs planning in the space of internal coordinates [39,44−46]. In this case, the controller should be additionally equipped with inverse kinematics unit and a unit for approximation or interpolation of the preset trajectory.

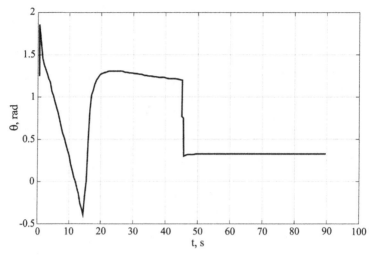

FIGURE 1.10 Vehicle's orientation angle.

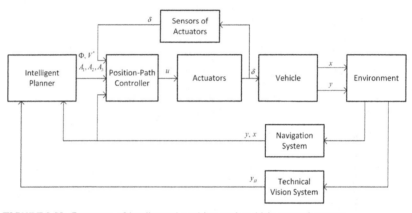

FIGURE 1.11 Structure of intellectual position-path vehicle control system.

The limitations on vehicle wheels' rotation speeds are set by the following inequalities:

$$\begin{aligned}
\text{if} |\omega_l| > \omega^{\max} \quad &\text{then} \quad \omega_l = \omega^{\max} \text{sign}(\omega_l), \\
\text{if} |\omega_r| > \omega^{\max} \quad &\text{then} \quad \omega_r = \omega^{\max} \text{sign}(\omega_r)
\end{aligned} \tag{1.18}$$

The planner task is to form the matrix coefficients Eq. (1.10) of quadratic and linear forms Eq. (1.9).

It is assumed that the control law in Eq. (1.13), Eq. (1.18) is implemented with a small sampling interval as given by

$$\Delta t \ll \frac{1}{T_1}. \tag{1.19}$$

Since the coordinates of the current goal point p_{1f}, p_{2f} are refreshed with a step of Δt, the even motion of a robot is ensured.

The following direct performance criteria are used for quality estimation [52]:

- Safety criterion (SC), S_m—minimal distance between any robot sensor and any obstacle along all the paths. This criterion determines the maximal risk during all the motion. Its weighting coefficient is taken to be equal to 20%.
- Path length (PL), P_L—length of all the paths traveled by the robot from the initial point to the goal with a weighting coefficient of 20%. For the trajectory set on the surface $p_1 p_2$ consisting of n node points assuming that the initial point has coordinates $(p_i, f(p_i))$ and goal coordinates $(p, f(p))$, the path length P_L can be calculated as follows:

$$\sum_{i=1}^{n-1} \sqrt{(p_{i+1} - p_i)^2 + (f(p_{i+1}) - f(p_i))^2} \tag{1.20}$$

$(p_i, f(p_i))$, is the trajectory point in Cartesian coordinates system, where $i = 1$, 2, ..., $n-1$.

- Task completion time (TCT), t_m—time required to complete motion with a weighting coefficient of 20%.
- Mission success coefficient (MSC), F—number of successful missions in undetermined environments with complicated obstacles with 40% weighting coefficient.

The cumulative effectiveness coefficient is proposed to be calculated by the following equation:

$$K = \sum_{i=1}^{n=4} a_n \cdot b_n \tag{1.21}$$

where a_n is the weighting coefficient and b_n the corresponding normalized value of n-th criterion.

Therefore the synthesis problem for intelligent planner of vehicle's motion can be formulated as follows.

A priori known coordinates of initial and final path points are defined in Cartesian coordinates system and are denoted as (p_1^0, p_2^0), (p_1^k, p_2^k), respectively. The vehicle is considered as an object with a lumped mass of a "material point" type with a tank-type chassis, with parameters such as current coordinates (p_1, p_2), yaw angle α, goal angle φ, liner course speed V_k, maximum permissible values of angular velocities of the robot wheels $(\omega_L^{max}, \omega_R^{max})$, wheel and chassis sizes, and a number of others.

The environment is not changing except for the obstacles that can be either static (stationary) or slowly moving on a surface. The motion speed is significantly (many times) lower than that of the vehicle. In addition, the single obstacles have a convex form. The obstacle can be grouped into certain configurations such as segments, lines, polygons, bays, and angles presented in Fig. 1.3. The uncertainty is caused by the fact that in application to a vehicle these obstacles are not localized. The control system has no preliminary information about their form, size, or position.

For situation assessment during the vehicle's motion, the vehicle is equipped with an onboard forward-looking locator with a limited range that forms a set of rays of various intensities having a space—time distribution. The dimension of the data set formed by the locator is determined by the beam angle from −45 to +45 degrees and range of 5 m.

The planner should form the coefficients Eq. (1.10) and manifold Eq. (1.9) matrices.

1.6 SUMMARY

In this chapter, a vehicle control systems' design problem analysis was performed. One of the most important problems is control design for nonlinear multiply connected systems, autonomously functioning under the conditions of parameters such as uncertainty and disturbances, and in the presence of stationary and mobile obstacles.

A vehicle's motion-model structure is considered and its simplification approaches are discussed.

A vehicle's path planning procedure satisfying the set goals is proposed. A general statement of a position-path control design task was made and a solution procedure is proposed. The position-path control laws are considered with simplifying assumptions in various motion models. Structural algorithmic solutions for the closed-loop systems are presented, and stability conditions for the planned paths are obtained.

At the controller level, other types of control laws can be used. If they are intended for planning motions in internal coordinates, the controller should include an inverse kinematics solver unit and a path approximation unit.

REFERENCES

[1] S.M. Alfimov, Prospective lines of basic military technologies development in the field of information control and processing systems implementation, Mechatronics, Automation and Control 2 (2008) 2—5.
[2] Yu.V. Chernuhin, V.Kh. Pshikhopov, S.N. Pisarenko, O.E. Trubachev, Sofware environment for modeling of the actions of adaptive mobile robots having two-level control system, Mechatronics 6 (2000) 26—30.
[3] N.I. Vaganov, Opening Speech at the Third All-Russian Research and Practice Conference—Perspective Systems and Control Problems, in: Proceedings of the Third All-Russian Research and Practice Conference, vol. 1, 2008, pp. 3—6. Taganrog.
[4] V.B. Sokolov, E.D. Teryaev, Unmanned Aerial Vehicles: Some Questions Concerning Development and Application, Mechatronics, Automation and Control, Moscow, vol. 2, 2008, pp. 12—23.
[5] F.L. Cernousjko, N.N. Bolotnik, V.G. Gradetskij, Mobile robots: Problems of Motion Control and Optimization, Proceedings of the XII All-Russian Conference for Control Problems (ARCCP), IPU of the Russian Academy of Science, Moscow, 2014.
[6] XII All-Russian Conference for Control Problems: Analytical Review, 2014. http://vspu2014.ipu.ru/taxonomy/term/101.

[7] V.Yu. Rutkovskij, Proceedings of the Control Problems Institute in the field of the reference model adaptive systems and astronautic aeroballistic control systems, Automation and Remote Control 6 (1999) 42—49.

[8] V.Yu. Rutkovskij, I.N. Krutova, Building-up Principle and Some Questions of One Class Theory for the Self-Adaptive System with a Model, Self-Adaptive Automated Systems: Proceedings of the I-st All-Unit Conference for Self-Adaptive Systems Theory and Practice (December 10—14, 1963), Moscow, 1965, pp. 46—63.

[9] V.Yu. Rutkovskij, V.N. Ssorin-Chaikov, Self-Adaptive Systems with Trial Input, Self-adaptive Automated Systems: Proceedings of the I-st All-Unit Conference for Self-adaptive Systems Theory and Practice (December 10—14, 1963), Moscow, 1965, pp. 93—111.

[10] S.D. Zemlyakov, Some problem of analytical synthesis in model reference control systems by the direct method of Lyapunov, in: Theory of self-adaptive control system: Proc. of Intern. Symposium, England, Teddington, 1965, P.H. Hummon Plenum Press, New York, 1966, pp. 175—179.

[11] B.N. Petrov, V.Yu Rutkovskij, S.D. Zemlyakov, Adaptive Coordinate-Parameter Control of Time-dependent Objects, Nauka, Moscow, 1980.

[12] I.M. Makarov, V.M. Lohin, S.V. Manko, M.P. Romanov, Intelligent Systems of Autonomous Mobile Objects Control, Mechatronics, Automation and Control, Moscow, vol. 2, 2008, pp. 6—11.

[13] I.M. Makarov, V.M. Lohin, S.V. Manko, M.P. Romanov, Possibilities and reality of information control and processing intelligent technologies application during creation samples of weapons and military equipment of the next generation, in: Proceeding Third All-Russian Research Practice Conference—Perspective Systems Control Problems, vol. 1, 2008, pp. 15—18. Taganrog.

[14] A.E. Gorodetskiy, A.A. Erofeev, Design Principles for Intelligent Control Systems for Mobile Plants, vol. 9, 1997, pp. 101—110.

[15] A.V. Timofeev, R.M. Yusupov, Intellectualization of automated control systems, in: Proceedings of the Russian Academy of Science, Technical Kibernetics, vol. 5, 1994, pp. 72—78.

[16] M. Vukobratovich, B. Karan, Algorithms of robot motion in the free space control, in: Proceedings of the Russian Academy of Sciences, Control theory and systems, vol. 1, 1995, pp. 205—220.

[17] V.V. Vojnov, A.S. Yushchenko, Control of a micro robot for medical purpose using fuzzy finite-state automaton and situation control methods, extreme robotics. Nano-, micro- and macro robots, in: Proceedings of the International Conference, Saint Petersburg: Polytechnica-Service, Gelendzhik, 2009, pp. 115—116.

[18] A.A. Krasovsky, Non-traditional target functionals and problems of optimal control theory (review), Proceedings of AN SSR, Technical Cybernetics 1 (1992) 3—15.

[19] A.A. Krasovsky, Development of generalized work minimum principle, Automation and Remote Control 1 (1987) 13—25.

[20] A.A. Krasovsky, Algorithmic basis of the optimum adaptive controls of the new class, Automation and Remote Control 9 (1995) 104—116.

[21] V.S. Lapshov, V.P. Noskov, I.V. Rubtsov, Experience in development and creation of semi-autonomous and autonomous special purpose mobile robot systems, in: Proceedings of TRTU, Special issue—Perspective Systems and Control Problems, vol. 3, 2006, pp. 35—40. Taganrog.

[22] V.M. Budanov, E.A. Devyanin, Concerning wheeled robots motion, Rational Mechanics and Mathematics 67 (2) (2003).

[23] S.F. Burdakov, I.V. Miroshnik, R.E. Stelmakov, Wheeled Robots Motion Control Systems, Saint Petersburg: Science, 2001.

[24] G.S. Byushgens, R.V. Studnev, Airplane Dynamics. Spatial Motion, Machine Engineering, Moscow, 1983.

[25] V.V. Gulyaev, O.F. Demchenko, N.N. Dolzhenkov, Mathematic Modeling during Forming Aircraft Image, Machine Engineering, Moscow, 2005.

[26] N.F. Krasnov, Aerodynamics: In Two Volumes, vol. 1, Machine Engineering, Moscow, 1976.

[27] V.Kh. Pshikhopov, Dirigible Balloons: Possibilities of Application in the Robotechnics, vol. 5, 2004, pp. 15−20.

[28] V.Kh. Pshikhopov, M.Yu. Medvedev, Mobile object Control in Definite and Indefinite Environments, Nauka, Moscow, 2011.

[29] V.Kh. Pshikhopov, M.Yu. Medvedev, A.R. Gajduk, Position-trajectory system of robotized aeronautical platform: mathematic model, Mechatronics, Automation and Control 6 (2013) 14−21.

[30] V.A. Shchepanovskij, G.I. Shchepanovskaya, Computer simulation of aerospace systems vol. 1, Nauka, Siberian Publishing Company of the Russian Academy of Science, Novosibirsk, 2000.

[31] V.Kh. Pshikhopov, M.Yu. Medvedev, R.V. Fedorenko, M.Yu. Sirotenko, V.A. Kostiukov, B.V. Gurenko, Aeronautic Complexes Control: Design Theory and Technology, Fizmatlit, Moscow, 2010.

[32] V.Kh. Pshikhopov, Repellers organization during mobile robot motion in the environment with obstacles, Mechatronics, Automation and Control 2 (2008) 34−41.

[33] V.I. Gurman, V.N. Kvokov, M.Yu. Uhin, Approximate methods of aircraft control optimization, Automation and Remote Control 4 (2008) 191−201.

[34] A.S. Devyatisylny, I.B. Kryzko, Stabilization of unmanned object motion along the programmed trajectory, in: Proceedings of the Russian Academy of Sciences, Control Theory and Systems, vol. 4, 1995, pp. 228−233.

[35] P.D. Krutko, Aircraft side motion control. Algorithms synthesis by inverse dynamic problems method, in: Proceedings of the Russian Academy of Sciences, Control Theory and Systems, vol. 4, 2000, pp. 143−164.

[36] P.D. Krutko, Aircraft longitudinal motion control, algorithms synthesis by inverse dynamic problems method, in: Proceedings of the Russian Academy of Sciences, Control Theory and Systems, vol. 6, 1997, pp. 62−79.

[37] E.M. Firsova, Adaptive system of maneuver aircraft control based on the division of the motion as to rolling and hunting, in: Proceedings of the Russian Academy of Sciences, Control Theory and Systems, vol. 1, 2001, pp. 110−119.

[38] M. Haouani, M. Saad, O. Akhrif, Flight control system design for commercial aircraft using neural networks, in: Proc. of the 15th Triennial World Congr. of IFAC (B'02), Barselona, 2002.

[39] Y. Kanayama, A. Yuta, M. Takada, J. Iijima, A locomotion module for wheeled mobile robots, Journal of the Robotics Society of Japan 2 (5) (1984) 402−416.

[40] Yu.G. Martynenko, Wheeled mobile robots motion control, Fundamental and Applied Mathematics 11 (8) (2005) 29−80.

[41] Y. Kanayama, Sh. Yuta, Vehicle path specification by a sequence straight lines, IEEE Journal of Robotics and Automation 4 (3) (1988) 265−276.

[42] T. Tsumura, N. Fujiwara, T. Shirakawa, M. Hashimoto, An experimental system for automatic guidance of roboted vehicle, following the route stored in memory, in: Proc. of the 11th Intern, Symp. on Industrial Robots, 1981, pp. 187−193.

[43] T. Hongo, H. Arakawa, G. Sugimoto, et al., An automatic guidance system of self-controlled vehicle—The Command System and Control Algorithm, in: Proc. IECON, 1985.

[44] V.V. Evgrafov, V.V. Pavlovskij, Dynamics, control and modelling of the robots with differential drive, in: Proceedings of the Russian Academy of Sciences, Control Theory and Systems, vol. 5, 2007, pp. 171−176.

[45] YuG. Martynenkob, A.M. Formalskij, Concerning omni-wheeled mobile robot motion, in: Proceedings of the Russian Academy of Sciences, Control Theory and Systems, vol. 6, 2007, pp. 142−149.

[46] J. Iijima, Y. Kanayama, S. Yuta, A locomotion control system for mobile robots, in: Proc. of the 7th Intern. Joint Conf. on AI, 1981, pp. 184−779.

[47] B.P. Demidovich, Lecture on Mathematic Theory of Stability, Nauka, Moscow, 1967.

[48] N.N. Krasovsky, Some Tasks of Motion Stability Theory, Fizmatlit, Moscow, 1959.

[49] N.N. Krasovsky, Problems of Controlled Motions Stabilization, Supplement to the Book of I.G. Malkin, Controlled Motions Theory, Nauka, Moscow, 1966, pp. 475−571.

[50] A.M. Lyapunov, General task on motion stability, Merkurij-Press, Cherepovets, 2000.
[51] V.D. Furasov, Motion Stability Estimation and Stabilization, Nauka, Moscow, 1966.
[52] G. Cielniak, A. Treptow, T. Duckett, Quantitative performance evaluation of a people tracking system on a mobile robot, in: Proc. of the Eur. Conf. on Mobile Robots (ECMR), Ancona, Italy, 2005.

Chapter Two

Neural Networking Path Planning Based on Neural-Like Structures

V. Guzik, A. Pyavchenko, V. Pereverzev, R. Saprykin, V. Pshikhopov

Southern Federal University, Taganrog, Russia

2.1 BIONIC APPROACH TO BUILDING A NEURAL NETWORK—BASED VEHICLE PATH PLANNER IN 2D SPACE

According to Refs. [1,2] the essence of the bionic approach for creation of intelligent systems is that a technical system imitates a biological one to demonstrate "intelligent" behavior in a complicated a priori non-formalized environment. This approach is based on discovering and using biological analogies. It particularly concentrates on using results of neurophysiological and neurocybernetic experiments on human and animal neural systems. At the same time, it can be admitted that direct application of the known neurophysiological data for human psyche modeling does not give any practically useful results due to extreme complexity of the modeling object, i.e., human brain.

As we know, human psyche is a system of highly organized matter with cognitive phenomena that can rarely be explained by the results of fine neurophysiological experiments. This is the reason for the difficulties faced in creating the general artificial intelligence systems, and particularly intelligent path planning systems.

At the same time we should account that various behavioral acts are performed not so much at a conscious level as at an unconscious one and are characteristic of not only human brain but the brain of much simpler organisms, such as mammals or even insects. This means that creation of intelligent path planning systems can be started with design of devices, modeling not all the functions of the natural highly developed intelligence, but imitating only those connected with performing purposeful actions and ensuring intelligent behavior in the natural environment. In other words, it would be wise to start modeling

25

Path Planning for Vehicles Operating in Uncertain 2D Environments. http://dx.doi.org/10.1016/B978-0-12-812305-8.00002-8

the behavior reflexes not on the psychological level but on a simpler neurophysiological one.

The researchers have advanced in understanding the structure and functions of the human nervous system that underlie the reflexive behavior. Particularly, it is known that any nervous system consists of a network of connected neurons functioning in parallel. Despite the comparatively low response speed of each neuron (several milliseconds), their simultaneous functioning allows the nervous system to process large bulks of information coming from multiple receptors and to form control actions for the actuators in real time.

Based on the fact that afferent synthesis and control by objectives are common grounds for control techniques on various levels of biological organization, we can assume that looping the feedback of a simplest homeostatic system by environment (Fig. 2.1) we can get a model for a basic intelligent motion planning system (an intelligent planner).

Such a system is essentially simpler than the functional system of behavioral level presented in Ref. [1] and includes a structure of afferent synthesis (SAS) implemented by means of artificial NNs together with a special decision-making scheme (DMS) located at the output (Fig. 2.2).

From Fig. 1.11, it follows that the inputs of SAS are the numerical values of distances to the closest obstacles generated by the technical vision system (TVS) and data coming from the vehicle's built-in navigation system. The TVS is capable of perceiving the environment as a collection of sections of three types: free, forbidden, and goal. We will also assume that TVS has such links with the afferent synthesis subsystem that a perfect correspondence

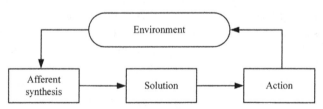

FIGURE 2.1 Functional intelligent system looped by environment.

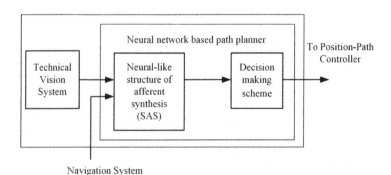

FIGURE 2.2 Neural networking planner. Two-component structure.

is established between the sections of the environment and elements of the NN of afferent synthesis that react according to the state of these sections. Since the TVS is rigidly bound to the vehicle's body, it would be natural to assume that the origin point of the coordinates system reflected in SAS is also fixed to the vehicle's body [2].

Let us describe the essence of the considered method. The vehicle is in the process of an active interaction with an a priori, non-formalized environment. Prior to any elementary action of the actuating subsystem, the NN consisting of artificial neurons retrieves the environment plan created in TVS that reflects the mutual location of the goal, obstacles, and the vehicle [3]. The NN is then used to find the functional gradient determining the set of possible paths reaching the goal. After that the path segment is generated—the one that is directed along the antigradient vector of the functional determined by the NN.

The considered way differs from all the known ones by the fact that the sought vehicle path is generated by means of physical modeling of the afferent synthesis process performed by the NN. The information about the goal and obstacles positions can be interpreted as circumstantial afferentation. The goal getting into the robot's field of perception or setting of this goal on a plan by any other way (e.g., by an operator through a communication channel) can be a model of starting afferentation. It is important that step fixation of the future action (path following) is performed not on the environment plan but in the state of the object itself. The high speeds of the TVS and SAS NNs make it possible for the built-in control system to react on emergence of dynamic obstacles and evade them safely and in-time.

2.2 SYNTHESIS OF NEURAL NETWORKING PLANNER AS A PART OF POSITION-PATH CONTROL SYSTEM. TASK STATEMENT

As mentioned in Chapter 1, a system of technical vision is used to estimate the situation in front of the vehicle. It is implemented in the form of a forward-looking multibeam locator. Let us simplify the task of getting and processing data coming from TVS assuming that instead of time—space intensity distribution the multibeam locator periodically generates a vector of distances to the detected obstacles. In this case, with some assumptions, a locator can be substituted with a laser radar (lidar) with limited range.

Let the lidar's visibility range be limited by the obstacle detection range value $0...D_{max}$ (in the test, up to 5 m), by the beam angle Ω (in the test, up to 91 degrees, beam width λ (in the test 1 degree) and quantity N equal to Ω/λ. Fig. 2.3 gives an example of a sector scan pattern for a multibeam lidar indicating a power center of each beam illuminating an obstacle.

Let us make an assumption that all obstacles have the same beam reflection ratio and there is no beam dissipation or absorption by the environment. In other words, we will consider that the distances determined are 100% accurate and the reflected signal is valid, without noise, and has no space-time distortions.

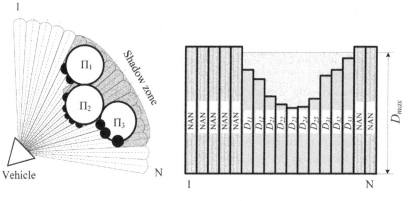

Example of lidar's angular pattern

Example of vector distance histogram, corresponding to the figure on the left

FIGURE 2.3 An approximate form of the beam pattern and generated distances vector.

If there is no reflected signal in the activated "beam channel" or the power of the reflected signal is very low (below threshold), the lidar generates a non-numerical value ∞ defined as NAN.

Let us also assume that the locator gives out the distances vector values to the intelligent system of position-path control (ISPPC) on demand. The data request is generated by the control system itself on a regular basis as the vehicle controlled by its actuating subsystem moves toward the predicted position with a preset precision ξ_r. Requests are performed one time per ISPPC working cycle.

Problem statement. The planner should perform real-time processing of the lidar data being a vector of distances to the detected obstacles. The planner knows the current vehicle's position and in proper time calculates and passes the direction parameters to the position-path controller so that it can safely continue its motion during the next time interval.

The safe motion is understood as one that avoids collisions of the vehicle with the detected obstacles.

According to Section 1.5, after each working cycle the designed planner calculates (Eqs. 1.9 and 1.10) and passes the values of the coefficients matrices A_{ij} and angle φ to the position-path controller. Additionally, the controller could get the following values: ξ, parameter that is equal to 0 (positioning task) or 1 (motion with a preset speed as in our case); the recommended value of the cruising speed V_k and the ceiling values for robot's wheels angular velocities. As the controller finishes calculations, ISPPC passes the new valid values of the Cartesian coordinates (p_1,p_2) of the new vehicle's position that became current.

The vehicle continues its motion until it reaches the goal point with coordinates $\left(p_1^k,p_2^k\right)$ inside the confidence limit with a preset accuracy equal to ξ_k.

Let us note that it is necessary to perform the planner development in a formal-logical basis of NN based on the idea of bionic approach to the creation of intelligent systems. The essence of this idea was described in Section 2.1. The planner effectiveness can be given for two cases. In the first one its kernel

Table 2.1 Initial Data for Neural Networking Path Planner Synthesis Task

Parameter	Value
Input data	Goal position, TVS data (laser radar), vehicles current position
Output data	Required vehicle's turning angle, recommended vehicle's speed (or predicted coordinates of the new vehicle position)
Dimension of the environment	2-Coordinate
Environment representation type	Cartesian
Functioning requirements	Real-time planning, reaching the goal in a limited time
Finding the obstacles	It is allowed to consider all obstacles as convex geometrical structures.
Forming the global/local paths	Possibility to move along the shortest path
Type	Built into vehicle's ISPPC
Formal logical basis of the planner's kernel	Feedforward neural-like networks

will be implemented as loosely coupled multilayer NN of a regular type [4,5], and as convolutional neural network (CNN) in the second one [6,7]. That is why the designed planner can be called a neural networking planner.

In a generalized form, we collected the initial data necessary for neural networking path planner (NNPP) synthesis as shown in Table 2.1. The NNPP is a part of ISPPC and its effectiveness can be estimated using the functioning performance criteria defined in Chapter 1.

2.3 DEVELOPMENT OF THE BASIC METHOD OF DETERMINING THE VEHICLE'S MOTION DIRECTION UNDER THE CONDITIONS OF UNCERTAINTY

2.3.1 Task Statement

On the basis of neural networking formal logic, let us find the solution of the planning task under the partial uncertainty condition during the vehicle's motion from the initial point to goal.

Taking into the account that vector representation of the distances coming from lidar is a particular case of processing the space-time representation of the intensities of the reflected signal generated by the real locator, first we will try to solve our task without using the discrete representation of the environment in the form of graphic grid with orthogonal or hexagonal link topology.

2.3.2 Method Development on a Classical Basis

We will develop a method based on so-called class of algorithms without memory.

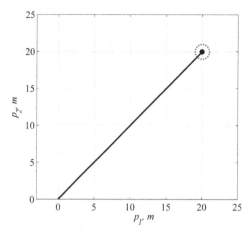

FIGURE 2.4 Example of straight motion of a vehicle from the point with coordinates (0; 0) to the goal point (20; 20).

Obviously, if the object during its motion will keep the direction toward the goal, and if there are no obstacles, we get a straight vehicle's motion along the shortest path (Fig. 2.4).

If there is an obstacle located directly on the straight path, it is necessary to deviate from the shortest path in order to avoid the collision. It would be necessary to determine in what direction and to what extent we should perform the maneuver.

Obviously, the collision probability is higher for higher speeds. In order to increase the vehicle's safety we need to reduce the vehicle's maximal allowed speed as it approaches an obstacle. The area where the speed is limited due to a possible collision is termed as risk area.

Using the idea of potential fields presented below, let us empirically define six risk zones with a recommended value of course speed V_k. Let us assign numbers to the zones in reverse proportion to the risk factor. The fifth zone (green) is a neutral zone without limitations and the motion speed is set to a maximum level of V_{max} here. The fourth zone (white) is a zone where the maximal speed is reduced by 1/5 and set to a value of $\frac{4}{5}V_{max}$. The third zone (yellow) is a zone where the speed is limited by $\frac{3}{5}V_{max}$. The second zone (orange II) is a zone with a speed limit of $\frac{2}{5}V_{max}$. The first zone (orange I) has a speed limit of $\frac{1}{5}V_{max}$. Finally, the zero zone (red) where the speed V_k is set to zero. In the latter case, the vehicle's motion is prohibited and the course angle α is set to a deflection of ϑ_α degrees to the right with respect to the current course angle (spot turn using the "right hand" rule). In the ith zone the linear speed of the vehicle can be determined as $\frac{N_z - i}{N_z}V_{max}$, where N_z is number of risk zones. Here $i = 0$ corresponds to the green zone and $i = N_z$ to the red one.

Since the lidar's input is limited to a sector, and there is no information about the obstacles location at the starting moment, the vehicle's control system will account for the danger zones dynamically as it receives the information about the obstacles.

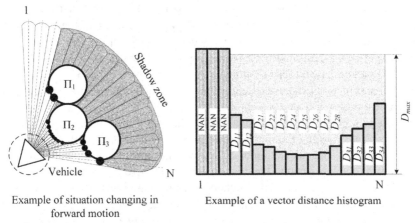

Example of situation changing in forward motion

Example of a vector distance histogram

FIGURE 2.5 Example of a model situation changing in forward motion and corresponding vector distance histogram.

The approximate distribution of the risk zones presented above will be used in modeling of the proposed approach. The zone size, on each new iteration, will be set in proportion to the value of the current cruising speed determined at the previous iteration. The higher the speed is, the higher becomes the risk level if the obstacle appears. Speed reduction results in lowering of the risk level. Such an approach allows to calculate a recommended vehicle motion speed for the next iteration depending on the environment state ahead. In addition, it makes it possible to closely approach the obstacle reducing the value of V_k and, therefore, the risk level, virtually to zero.

In order to imitate the potential filed, it is not sufficient to reduce the acquired numerical values of the distances by the calculated value proportional to the motion speed. It is also necessary to perform angular widening of the obstacle images illuminated by the locator. Let us note that the motionless obstacle's angular size increases as we approach it. The obstacle increasingly covers the active zone of a locator (scanned area) (see Figs. 2.3 and 2.5). The width of such zone is selected in a way that the object moving along a tangent line would not go by the edge of the obstacle (which is typical of most of the classical shortest path search algorithms) but in a distance sufficient for collision avoidance.

As a remark, let us note that in a natural experiment selection of the risk zones and new recommended value of the course speed is, to a large extent, dictated by the values of the current speed, the power and mass characteristics of the vehicle, and environmental factors.

In order to simplify the task under the conditions of limitations put on linear and angular velocities of the vehicle, let us select the value of angular widening of the obstacles to be constant and a multiple of angular width of a separate beam of the locator.

The path planning method proposed below, just as vector field histogram (VFH) method [8], is based on presenting the sonar data in the form of vector histograms of numerical and non-numerical values. The mentioned method implies building a grid representation around the vehicle, and the histograms illustrate the distribution of (intensity) energy density values of the reflected signal binding it to the all-round (360 degrees) grid. Unlike this method, our basic method variant has the following features:

1. There is no space discretization by the grid method and therefore there is no problem of a grid step selection.
2. The histograms used in calculations are the vector distribution of distances to obstacles in the limited sector of frontal field of view with an angular width of Ω. The calculation of distances to the detected obstacles is moved outside the basic method and is not considered here.

Such an approach is called a method of vector histograms or distance vector histogram (DVH) method. It includes the following sequence of actions:

1. Determine the initial parameters of the experiment including obstacle location, coordinates of the vehicle's initial position and of the goal.
2. In the basic system of Cartesian coordinates, we calculate the initial values of goal angle and of the course angle φ_0 with respect to the initial position of the vehicle.
3. Set the value of the maximum vehicle motion speed V_{max}.
4. Determine the initial recommended value of course speed and declare it for current (V_k).
5. Set the initial distribution for the corresponding risk areas, the recommended linear vehicle motion speeds, limitations on the angular velocities of the robot wheels, and the required values of parameters of the controller as a part of its ISPPC.
6. Set the angular widening of the obstacles (number of locator beams by which the obstacle image widens once it is caught into the active zone of the locator mounted onboard the vehicle). For a particular case, it can be selected to be constant without any relation to the remaining distance to the obstacle.
7. Activate the TVS.
8. Depending on the current recommended linear motion speed V_k, we perform calculation of the corresponding values of actual dimensions of the risk areas.
9. Form a request to the TVS (lidar in our case) and get a vector of distances to the possible obstacles.
10. Form a vector of empty elements called a course vector with dimension equal to the dimension of the distances vector generated by the locator. The middle element of the course vector will correspond to a 0 course deviation in a next motion step (no course change). The initial element corresponds to deviation by $\frac{\Omega}{2}$ to the left (positive deviation angle); the final element corresponds to deviation by $\frac{\Omega}{2}$ to the right (negative deviation angle).
11. With respect to vehicle's current course angle φ_i, we calculate the value of apparent angle of course deviation $\tilde{\alpha}$ corresponding to the desired direction toward the goal (Fig. 2.6). If the value of the apparent angle of course

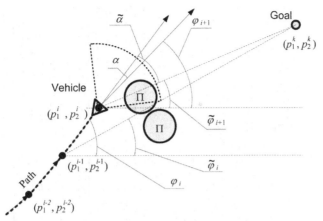

FIGURE 2.6 Geometric representation of the obstacle avoiding parameters.

deviation $\tilde{\alpha}$ lies within $\pm\frac{\Omega}{2}$, we select the element index in the course vector corresponding to the goal direction. In case when value of $\tilde{\alpha}$ is not in the mentioned range, index is set to initial element of the course vector for $\tilde{\alpha} > +\frac{\Omega}{2}$, or to the final one for $\tilde{\alpha} < -\frac{\Omega}{2}$.

In our experiment, the course angle φ_i is as an angle between the abscissa and the line passing through two points (p_1^{i-1}, p_2^{i-1}) and (p_1^i, p_2^i)—previous and current vehicle positions of $(i-1)$th and ith iterations, respectively. Except for the initial moment when the course angle φ_0 is equal to the slope of the line passing through the points with coordinates (p_1^0, p_2^0) and (p_1^k, p_2^k). Then we put the goal label into the selected element of the course vector. Fig. 2.6 introduces the following notation: φ is the value of true course angle, $\tilde{\varphi}$, value of the apparent course angle, α, value of the true deviation angle, and $\tilde{\alpha}$, value of the apparent deviation angle.

12. Using the distances vector obtained from the locator for all the numerical values we determine the distance to the closest of the detected obstacles, i.e., we determine an element (elements) with their value(s) being minimal at the current step (algorithm iteration).

Perform an angular widening of the convex obstacle edges. Generally, it is performed along the vector to the left and to the right of the found minimal elements by n_λ elements, i.e., by $n_\lambda \cdot \lambda$ degrees (see Fig. 2.7).

The number n_λ is set a priori (step 1 of the algorithm). Thus, in the experiments for the set scene templates the value of constant n_λ was selected to be equal to 4 (varied from 3 to 7). In realization attention should be paid to the following:

a. we expand the initial edges of the obstacles, not the modified ones; and

b. value of each element of the distance vector is determined as a function of search for a minimum between the old value of this element and values obtained for this element in performing the widening operation.

If the search reveals several identical minimal elements, the last to be widened will be the one that is located closer to the goal. This reduces the number of distance vector elements holding non-numerical values

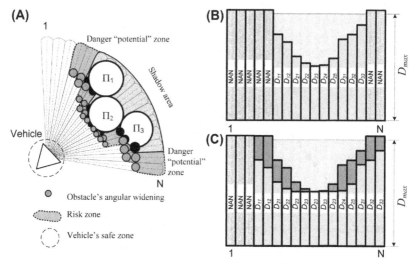

FIGURE 2.7 Example of a model situation changing (A). Distance distribution before (B) and after (C) the angular widening of the obstacles edges.

NAN. An exception is the case when locator detected no obstacles, then the distances vector is not changed.

13. In the distance vector all the NAN elements are assigned the value $D_{max} + \delta$, where δ is some numerical value that allows substitute non-numerical value NAN with a numerical value knowingly greater than D_{max}, maximal detection range of the locator (in the experiment $\delta = 0.02$). For the rest of the elements of the distance vector we perform a subtraction operation on the values of the actual risk zone size corresponding to the risk degree assigned in the previous iteration. The obtained distances will be called *apparent*. As a result, a risk zone of an actual size will be located between the moving vehicle and the obstacle, similar to the one presented in Fig. 2.8A.

In the simplest case the risk zone size value can be found as $V_k \Delta t_T$, where Δt_T is the vehicle deceleration time, i.e., reduction of the linear speed from recommended V_k to 0. The actual size of the risk zone cannot be negative. If the vehicle is not supposed to approach the obstacle closely, the size of the red zone should not go below a certain constant value set a priori (determined during experiments).

The mentioned correction is performed if the minimal element value found in the entire vector is greater than the actual risk zone size at the present moment.

As a result of accounting for the actual risk zone sizes, the distances vector will take a form similar to the one presented in Fig. 2.8C. Here the updated vector is backgrounded by the old values obtained at the stage of angular widening (see Fig. 2.7).

Fig. 2.8A shows that the proposed variant of the control zone setting corresponds to the predicted change of the scan sector for the vehicle

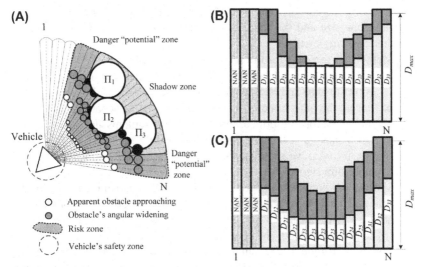

FIGURE 2.8 Example of a model situation changing (A) and distances distribution before (B) and after (C) the operation of accounting for the risk zones actual sizes.

approaching the obstacle at the distance of the risk zone actual size (compare Figs. 2.3, 2.5 and 2.8).

14. We search for a maximum among all the measured elements. If one maximum is obtained, the solution is found. If there are several solutions (we found several local maximums with equal amplitudes), then we select the value of the local maximum closest to the goal label on the right using the course vector values.

 The index of the distance vector elements so obtained is the sought heading (H). The corresponding local maximum will be called a course distances maximum (CDM).

 For the purpose of local maximums identification, we introduce a vehicle motion permission indicator (VMPI). The value of this indicator is set as "*true*" if at least one of the local maximums is found to be equal to $D_{\max} + \delta$ during the search process or the vector includes at least one pair of neighboring elements with their value difference exceeding a certain setting η. Otherwise VMPI is set as "*false*". The numerical value of η is selected in proportion to the vehicle's size. There is no sense to move in a direction where there is not enough space for motion. In the experiment, the vehicle was considered to be a particle and the minimal radius r of single obstacles was known. So in order to simplify the algorithm research procedure the value η was set to $0.9r$ as it was done in Ref. [9].

15. Using the freshly updated distances vector we determine the minimal apparent distance (ADM) to the obstacle. In order to set the robot's behavior rules for the red risk zone we introduce an indicator "right hand indicator" (RHI). We set this indicator to *false*.

16. Knowing the current values of ADM and CDM we determine the actual risk zone size and the value of the linear motion speed V_k using one of the following methods.

    Method 1. Accounts for the actual size of the risk zone.

    a. If we are in the red zone (or if VMPI is *false*) the motion speed should be set to zero and RHI to *true*. Go to the next algorithm step. Else:

    b. For all the other zones we select the value of V_k according to the recommendations described above. If the extremums get into different risk zones V_k is selected for the zone with a higher risk [9]. If there are no risks, we set V_k to be equal to V_{max} and continue to the next algorithm step.

    Method 2 is applied if the goal is in the line of sight (in the active locator zone), is not covered with any obstacles, and if the current robot's coordinates $\left(p_1^i, p_2^i\right)$ fall into a predefined vicinity of the goal with coordinates $\left(p_1^k \pm \Delta p_1, p_2^k \pm \Delta p_2\right)$. In the first case the motion step is calculated similar to method 1, excluding the option of an obstacle present directly ahead (i.e., V_k cannot take a zero value). In the second case V_k is set proportionally to the value half the distance to the goal along the straight line, but not more than V_{max}.

    Likewise, if necessary, we set the maximum values of angular velocities of the wheels for a vehicle that has a "tank" design.

17. Check the RHI state. If the indicator is *true*, we select a new recommended value of the course angle φ_{i+1} as follows

$$\varphi_{i+1} = \varphi_i + \vartheta_\alpha \qquad (2.1)$$

    where the increment ϑ_α is generally determined based on the relations of the angular velocities of the left and right wheels of a "tank" design vehicle for a zero value of V_k and spot rotation of the vehicle about the chassis center (see Section 2.1). So the resulting value ϑ_α should be less than zero.

    Proceed to the next (18) algorithm step.

    If RHI is *false*, in order to avoid the obstacle we use the previous H value to derive angular deflection from the current vehicles course angle α (denoted as α_H). The course angle is given by the current value of φ_i. The new recommended value of the course angle φ_{i+1} is given by:

$$\varphi_{i+1} = \varphi_i + \alpha_H \qquad (2.2)$$

18. Then knowing the coordinates $\left(p_1^i, p_2^i\right)$ and the values of V_k and φ_{i+1} we can calculate the coefficients matrices A_{ij} (see Eqs. 1.9 and 1.10).

19. The data obtained from calculation is sent to the position-path controller ISPPC. If required, the controller can receive the recommended values of maximum wheel angular velocities necessary for the next iteration.

    As a result under the action of position-path controller being a part of ISPPC the vehicle should move toward the vicinity of the preset point

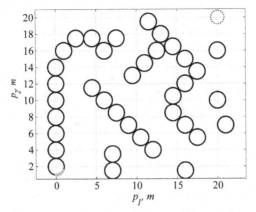

FIGURE 2.9 Scene 6—flat structured environment with obstacles.

with an absolute error less than ξ_r (0.02) in the experiment. This will give the new true values of the vehicle's position coordinates.

20. The actions are repeated in iterations starting from step 8 until one of the following occurs: either current vehicle coordinates coincide with the goal coordinates with a preset error ξ_k, or the path search algorithm gets into a local minimum exceeding a preset modeling time (e.g., exceeds the number of algorithm iterations).

The following section presents the results of modeling performed in MatLab.

2.3.3 Method Modeling Results in MatLab

According to the task statement presented in Chapter 1, we limit the experiment by 6 scenes (the basic scenes 1, 2, 3, 4, and 5 are presented in Fig. 1.3, and the additional scene 6 in Fig. 2.9). They include distributed configurations for solitary convex obstacles, L-shaped, and U-shaped obstacles (see Fig. 1.3).

For each basic scene we determine two pairs of coordinates:

$\left(p_1^0, p_2^0\right)$ are the coordinates of the vehicle's motion starting point equal to (0.0; 0.0);

$\left(p_1^k, p_2^k\right)$, vehicle goal coordinates given by the pair (20.0; 20.0).

The additional scene 6 is intended to check the method effectiveness in detail. For this reason we expand the initial conditions of the experiment:

- values of $\left(p_1^0, p_2^0\right)$ are set from the following ordered set {(20.0; 20.0); (0.0; 0.0); (0.0; 21.0); (20.0; 0.0); (20.0; 20.0)};
- values of $\left(p_1^k, p_2^k\right)$ from the ordered set {(14.0; 14.5); (20.0; 20.0); (20.0; 0.0); (0.0; 20.0); (0.0; 0.0)}.

Let us set the limitations on the linear course speed (mps) (see previous section). The recommended value of the motion speed in the green zone is determined as $V_{\max}$ equal to 1 mps.

Let us set analogous limitations on the angular velocity of the wheels of the robot's "tank" design (deg/s) (see previous section). The maximum value

$\omega^{\text{max}} = 25.0$; for the "red risk zone" and for the "orange risk zone (I)" the recommended value of speed $\omega^{\text{min}} = 0.2\omega^{\text{max}}$; for the "orange risk zone (II)" the recommended value is $0.4\omega^{\text{max}}$; for the "yellow risk zone" the recommended value is $0.6\omega^{\text{max}}$; and for the "white risk zone" the recommended value is $0.8\omega^{\text{max}}$. The recommended value in the green zone are defined as ω^{max}.

Radius r of each of the two robot's wheels is equal to 0.2 m. Half length of the chassis is equal to 0.5 m and the lumped mass center is located over the middle of the robot's chassis. This means that if $\{|\omega_l|; |\omega_r|\} \leq \omega^{\text{max}}$, the chassis can rotate on a surface by not more than $\vartheta_\alpha = 10$ degrees in 1 s.

Let the locator (lidar) range be equal to 5 m, and the value of δ be equal to 0.02 m. The values exceeding the maximum possible will be equal to 5.02 m. The other lidar parameters are $\Omega = 91$ degrees, $\lambda = 1$ degree, and $N = 91$.

Let us set the angular widening of the obstacles n_λ equal to 4 which correspond to 4 degrees. The value of η is set to 0.9 m.

Values of Δp_1, Δp_2 are set to 0.02 m. As the vehicle moves from current coordinates to the predicted ones, it has a positioning error of ξ_r set to 0.02 m.

Fig. 2.10 presents a generalized flowchart for the MatLab implementation of the presented DVH method.

Let us consider a number of key features for such an implementation.

In modeling procedure the value of the initial vehicle turning toward the goal was determined in a classical way based on the values of the corresponding coordinates of the initial vehicle $\left(p_1^0, p_2^0 \right)$ and goal $\left(p_1^k, p_2^k \right)$ positions:

$$\tilde{\varphi}_0 = arctg\left(\frac{p_2^k - p_2^0}{p_1^k - p_1^0} \right)$$

Analogously for each ith iteration of the algorithm, an angle of desired vehicle's direction toward the goal was determined using the corresponding coordinates—robot's current position $\left(p_1^i, p_2^i \right)$ and its goal $\left(p_1^k, p_2^k \right)$:

$$\tilde{\varphi}_i = arctg\left(\frac{p_2^k - p_2^i}{p_1^k - p_1^i} \right).$$

In calculations the angle was transformed to a positive value.

The shortest distance to the goal from the current position without accounting for the obstacles was calculated using the following formula:

$$L_{\text{min}} = \sqrt{\left(p_2^k - p_2^i \right)^2 + \left(p_1^k - p_1^i \right)^2}.$$

Recalculation of the risk zones was done at the beginning of each new iteration, with respect to the recommended course speed determined at the previous cycle. In the experiment we found two controlled sizes of risk zones that allow accounting for the introduced six risk levels (from green to red).

The minimal distances to the obstacles were found using the actual distances vector obtained from the lidar.

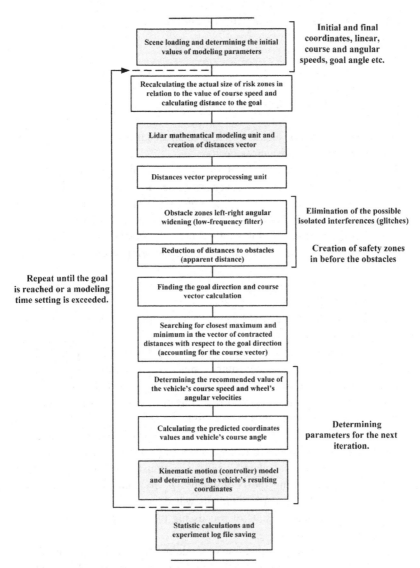

FIGURE 2.10 Algorithmic structure of the method model.

The angular widening of the obstacles was performed according to the step 12 of the algorithm using the idea of iterational low-frequency filtration of the distances vector implemented by the following actions.

Angular widening by n_λ elements to the right,

$j = 1{:}n_\lambda$

$\{i = 1{:}N.$

$\{D_{2,i} = D_{1,i}\}$

$i = 1{:}(N-1)$

$\{D_{1,i+1} = \min(D_{2,i}, D_{1,i+1})\}$

$\};$

Angular widening by n_λ elements to the left,
$j = 1:n_\lambda$
{ i = 1:N.
 { $D_{2,i} = D_{1,i}$ }
 i = 2:N.
 { $D_{1,i-1} = \min (D_{2,i}, D_{1,i-1})$ }
};

where $D_{1,i}$ is the ith element of the main distances vector and $D_{2,i}$, ith element of the auxiliary distances vector.

Due to such implementation of the step 12 of the algorithm, the non-numerical values NAN were replaced by the numerical values equal to $D_{\max} + \delta$ before this step, not after it as presented in the algorithm (step 13).

The presented method of angular broadening of the obstacles plays the role of a low-frequency filter. In this way, all the solitary positive glitches of the numerical values will be smoothed by the neighboring vector elements and, therefore, excluded from further analysis.

Accounting for the current size of the risk zone is performed according to the step 13 of the algorithm.

The values of CDM and ADM were found according to the steps 14 and 15 of the algorithm.

The recommended value of the vehicle's linear course speed (V_k) was found according to step 16 of the algorithm using the capabilities of conditional branching operators of the m-language by analysis of current values of CDM and ADM, using the values of actual risk zones sizes and the shortest distance to the goal from the robot's current position.

According to step 17 the new recommended value of the course angle φ_{i+1} was found in relation to the RHI state according to the expressions (2.1) or (2.2). In addition, the calculated angle was transformed to fit the range [0−360] degrees. Knowing the coordinates (p_1^i, p_2^i), and the values V_k and φ_{i+1} calculation of the A_i matrix elements was done.

Fig. 2.11 presents the results of processing the nine algorithm steps for a vehicle moving in scene 1.

Here you can see the visualization of modeling results such as overall view of the scene and motion path Fig. 2.11A; lidar active zone Fig. 2.11B; visualization of the corrected distances vector, course vector and selected robot's motion direction Fig. 2.11C. As seen in Fig. 2.11C, since the path on the right is closed for the robot, the planner took a decision to bypass the obstacle on the left keeping the minimal safe distance away from it.

Fig. 2.12 presents the MatLab modeling results for a vehicle moving in test scenes using DVH planning method. The experiments results presented in Fig. 2.13 confirm the method's efficiency in a maze-type scene 6 for various initial conditions of the experiment.

Despite the clear success of the experiments, it should be mentioned that further research has revealed the fact that the process can get into circularity if DVH method gets into a local minimum. The mission success estimates

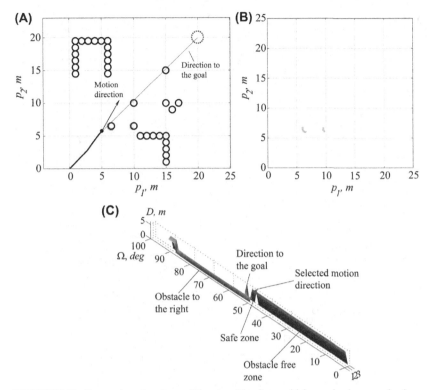

FIGURE 2.11 Processing nine steps of the algorithm for a vehicle moving in scene 1, where (A) overall view of the scene and motion path; (B) lidar active zone; (C) visualization of the corrected distances vector, course vector, and selected robot's motion direction.

presented in Fig. 2.2 cannot be considered as absolute. So Fig. 2.14 presents an example of situation when vehicles mission cannot be accomplished because the process gets into an infinite loop.

As the figure shows, the deadlock is caused by adding an additional single obstacle that blocks the robot's view as it goes through the rightmost lower point of the bay in the direction from right to left. Therefore, the planner selects the local maximum of the distances that lie outside the inner area of the U-shaped obstacle. That is why it misses the turn to the obstacle. This situation can be easily resolved by reducing the locator range in a repeated passing of the same scene area (see Fig. 2.15)

However, the necessity to perform a coordinate-wise check for a second pass will not only allow introduction of a special memory for keeping the previously passed coordinates but will also burden the method with time-consuming procedures necessary for comparing the current coordinates with the passed ones. The processing time would increase with the length of the path already passed. Such a check would expel this method out of the class of memoryless search algorithms and would require an additional research to make it balanced. Therefore, we do not present here a full solution for the problem of the vehicle getting into a local minimum.

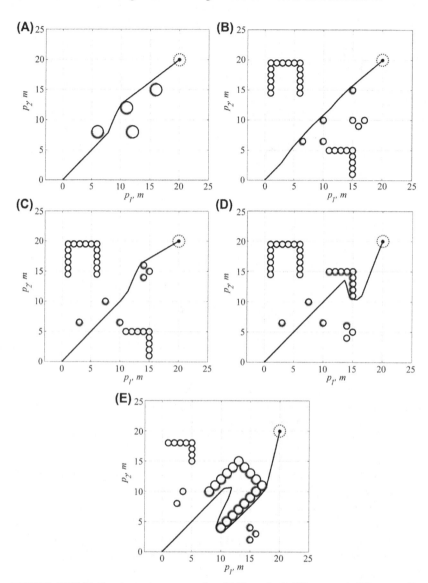

FIGURE 2.12 Motion in test scenes using DVH method: (A) scene 1; (B) scene 2; (C) scene 3; (D) scene 4; (E) scene 5.

From our point of view, the main advantages of the proposed DVH method are its simplicity and clearness, small computational costs, possibility to control vehicle's motion speed when it goes through risk zones, and usage of initial data "as is", i.e., in the form of distances vector (histogram) that is cyclically formed and fed by the locator to the planner.

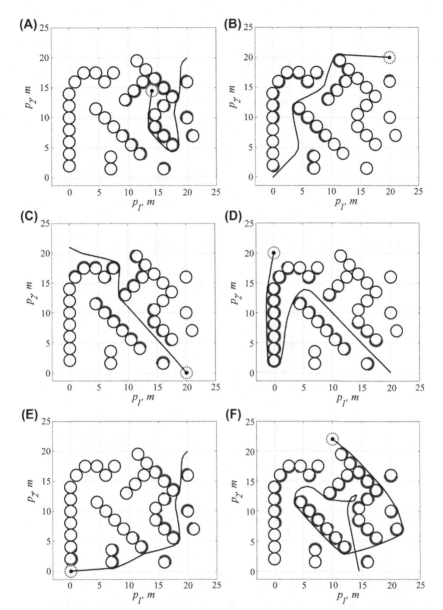

FIGURE 2.13 Motion in test scene 6 using DVH method: (A) path from (20.0; 20.0) to (14.0; 14.5); (B) path from (0.0; 0.0) to (20.0; 20.0); (C) path from (0.0; 21.0) to (20.0; 0.0); (D) path from (20.0; 0.0) to (0.0; 20.0); (E) path from (20.0; 20.0) to (0.0; 0.0); (F) path from (14.5; 0.0) to (10.0; 22.0).

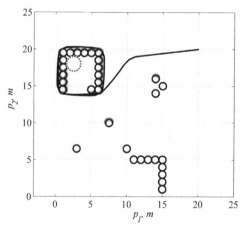

FIGURE 2.14 An example of the search procedure looping for a range equal to 5 m.

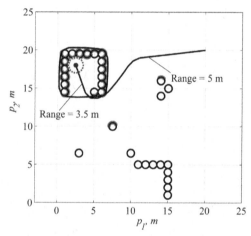

FIGURE 2.15 Example of solving the deadlock situation by reducing the locator range from 5 to 3.5 m in the second pass.

The essential drawbacks of the method are:

- possibility of getting into an infinite loop, which is typical of all the class of memoryless algorithms;
- low value of safety criterion, and its direct relation to the value of obstacle's angular widening;
- dependence of planner decisions' effectiveness on the accuracy of vehicle's current coordinates and goal coordinates detection in the process of vehicle's motion; and
- impossibility to predict the motion path several steps ahead.

It should be mentioned that the algorithm's quality is directly affected by the complexity of the scene, locator's range, and robot's motion speed.

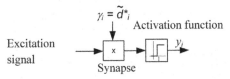

FIGURE 2.16 Structure of a neural-like element of DVH planner.

2.3.4 Implementation of DVH Method in Formal Neurons

A periodically updated histogram is going to be used as a sensory perception model. It should be mentioned that maximal amplitude values of the distances are limited by the capabilities of the used locator and by its preset range.

In order to implement the DVH method in a neural-like structure [10,11] each element of the distance vector is associated with a neural-like element the structure of which is presented in Fig. 2.16. Unlike the classical notion, a formal neuron (FN) is defined as a group of two components: a scheme (synapse) multiplying a single excitation signal by the value of synaptic coefficient and an activation function of a threshold type with a unity bias.

Number of such FNs is equal to the number of distance vector elements. The value of the synaptic coefficients γ_i in each ith FN is set proportional to the *normalized* value of the ith element of the distance vector. The distance vector itself should have undergone the operations of angular widening and accounting for the active risk zone.

The normalization operation for an ith element of corrected distances is understood as a ratio of the current value of the ith element to the current maximum, among all the elements of this vector:

$$\widetilde{d}_i^{\,*} = \frac{d_i}{\max\limits_{j\,=\,\overline{1,N}}(d_j)}. \tag{2.3}$$

Thus, if we feed all FNs of the line with an excitation signal, only the FNs that received unity synaptic coefficients corresponding to maximal normalized distance value in the current sample will be activated. Outputs of the other FNs of the line will remain in a passive (zero) state. Examples of the synaptic coefficients samples are presented in Fig. 2.17A and B. As these figures show, there can be a multitude of unity synaptic coefficients in a sample. Obviously, each unity value corresponds to a value of the course angle of possible vehicle motion, and generally only one of these values is true.

Similar to the Section 2.3.2, the actual course angle will be selected out of the found ones knowing the current goal angle (a direction closest to the goal is selected). For this reason the outputs of the considered FN line are connected to

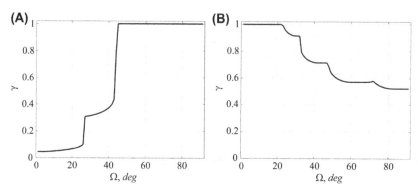

FIGURE 2.17 Examples of normalized distance distribution: (A) maximal distances values are in the center and to the right; (B) maximal distances values are at the left.

the signal inputs of a priority encoder. Each signal input of such a scheme is assigned the following:

- index (weight) having the same value as the output index of the corresponding FN; and
- priority level with its value being in inverse proportion to the difference between the course angle and the angle of the direction to the goal.

It should be mentioned here that the encoder inputs' priority level distribution should change according to the changes of the direction angle as the vehicle keeps moving. The maximum value of such distribution should shift proportionally and according to the direction toward the goal. As is well known, it is the Gaussian distribution (normal distribution law) that has the necessary qualities:

$$Koef(x) = \frac{1}{\sigma \cdot \sqrt{2\pi}} e^{-\frac{(x-m)^2}{2 \cdot \sigma^2}} \tag{2.4}$$

where m is the value of the function argument delivering maximum of the function.

Fig. 2.18A and B give several variants of normal distributions of priority levels of the encoder's signal inputs obtained for various directions to the goal (see parameter m in the expression 2.4) and accounting for the limitations of the method and for the case of coinciding old motion direction with the central element of the vector (middle of the locator's angular pattern 46 degrees in our case) (Fig. 2.19).

In order to avoid the uncertainty, each signal input of the encoder should be assigned not only a dynamic priority level but also a weight. This means that output of the encoder would give only one numerical value corresponding to the highest priority signal from an FN with the greatest weight. And this numerical value (weight) will be the sought shifted value of the actual course deviation angle.

Fig. 2.20 presents the structure of the neural-like unit for course angle calculation (NUCAC) for a moving vehicle. Let us present a brief functional description of this scheme.

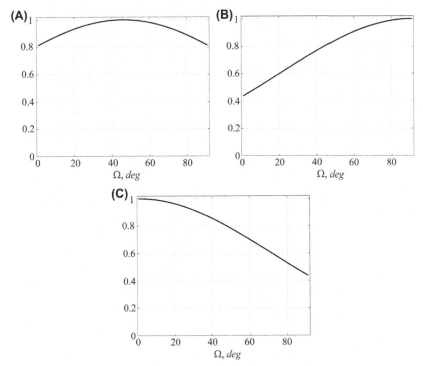

FIGURE 2.18 Examples of normalized function values: (A) goal straight ahead (shift 46 degrees); (B) goal on the right (shift 91 degrees); (C) goal on the left (shift 0 degree).

Under the action of activation signal using the values of current vehicle and goal coordinates, the angle of direction toward the goal is calculated and Gaussian distribution of priority codes is generated. The obtained values are then fed to the priority encoder where the ith value of the current priority code is assigned to the ith signal input of the encoder. After that the process of analysis of normalized distances vector elements is launched (activated). As a result of this analysis a number of FN fire in the line. The capture scheme located at the output of the line ensures capturing and holding the unity values of the corresponding activation functions in order to suppress the race effect and to avoid instability in the work of the encoder in the current step. The encoder generates the shifted value of the necessary course angle deviation value and feeds it to the unit of formation of the recommended course angle. The value of the current course angle is added here to the value of the required course deviation minus the preset shift value. The "captured" signal accompanies the formed value of the course angle ready for position-path controller action. After the controller processes the new course, unit is restarted.

The full structural scheme of the DVH planner built using NUCAC is presented in Fig. 2.21.

This structure supplements the capabilities of the course angle calculation unit by supporting reading and preliminary processing stages of the distance

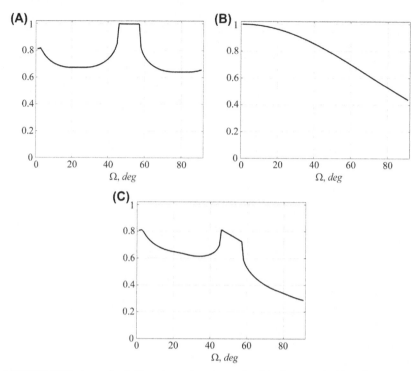

FIGURE 2.19 A result of distance vector processing for the set priorities distribution: (A) distances distribution; (B) normal Gaussian distribution of the priority function values; (C) two local maximums as a result of multiplication of the normalized values of distances by normalized values of the priority functions.

vector values. In addition, it handles the exceptions related to the formation of vehicle's motion inhibit (clearing the VMPI) and setting of RHI. It also supports the stages of calculation of linear and angular velocities values of the robot's wheels accounting for the actual size of the current risk zone for the obstacles detected ahead. The predicted values of robots coordinates are calculated and fed to the controller together with the value of the actual course for future processing. Activation of the planner denoted as DVH-NN is performed periodically in connection to the functioning of the position-path controller.

2.3.5 Modeling of DVH-NN Planner in MatLab

Most of the initial data for modeling is presented in Section 2.3.3. In addition calculation of Gaussian coefficients is performed for various goals. Preliminary calculation of coefficients is done for each value of one degree course angle shift (from 1 degree to 91 degrees). As a result we get a quadratic matrix with each row corresponding to the next value of goal's angular shift. Each element in its column contains coefficients of the Gaussian distribution.

The method's implementation is based on the approach described in Section 2.3.3, added by an operation of normalizing the values of the preprocessed

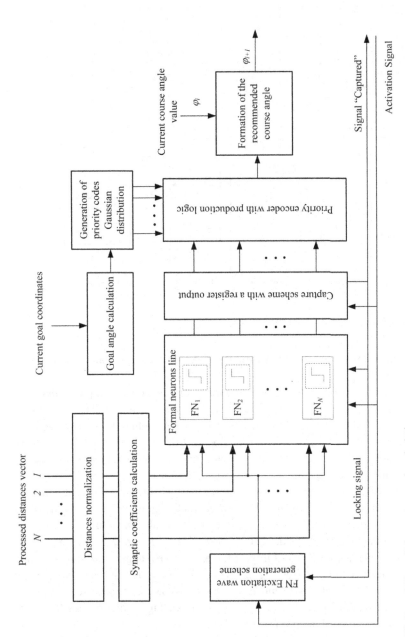

FIGURE 2.20 Neural-like unit for course angle calculation.

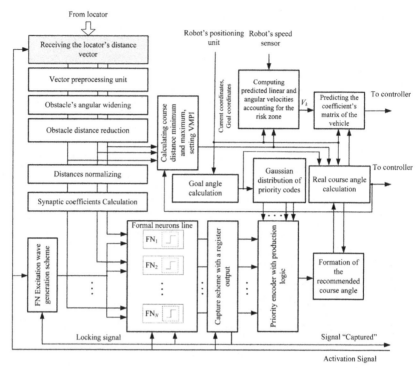

FIGURE 2.21 The DVH-NN planner structure.

distances vector and a software model of the course angle calculation neural-like unit.

In the software version of this unit the priority encoder is implemented with simplifications: each ith normalized distance value is multiplied by the value of the ith coefficient from the matrix row corresponding to the set goal. A maximum value closest to the direction to the goal is then obtained out of the results of vector product. If there are several equidistant values, the right-most one is selected (the value with a greater index).

However, the experiments showed that the substitution of priority encoder with a signal latching by an operation of multiplication of normalized values with a logical postprocessing, on one hand, results in a higher sensitivity when finding a path inside U-type obstacles (see Fig. 2.22), on the other hand, it increases the probability of the error course angle deviation selection for obstacle avoidance (see Fig. 2.22D and E). Nevertheless, since the method excludes the possibility to move in the red risk zone, the selected simplified software implementation can be used for modeling in order to estimate the method in the set configurational scenes.

Thus, on each iterational cycle of the software all FNs of the line are excited virtually simultaneously (due to consecutive software implementation). The maximal value of the activation function will indicate the winning FN. If several FNs win simultaneously, a neural element closest to the goal is selected

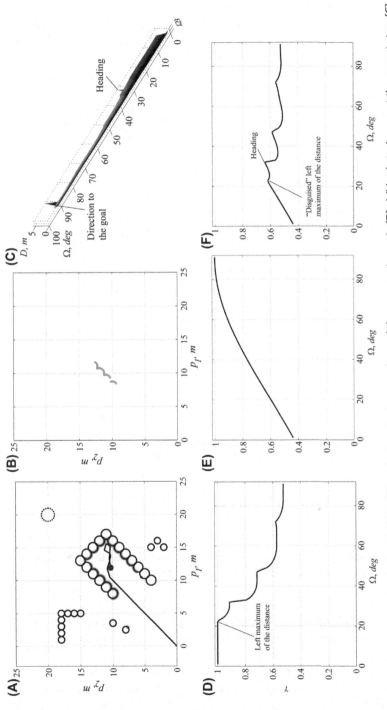

FIGURE 2.22 Interim results obtained in DVH-NN method application experiment: (A) scene passing path; (B) visible obstacle zone at the current step; (C) 3D histogram of the distances vector values with superimposed goal designation and heading marker; (D) 2D histogram of normalized distance vector values; (E) normal Gaussian distribution of priorities function values (goal on the right); (F) results of multiplication of normalized values of the distances vector by the normalized values of coefficients (priority levels codes).

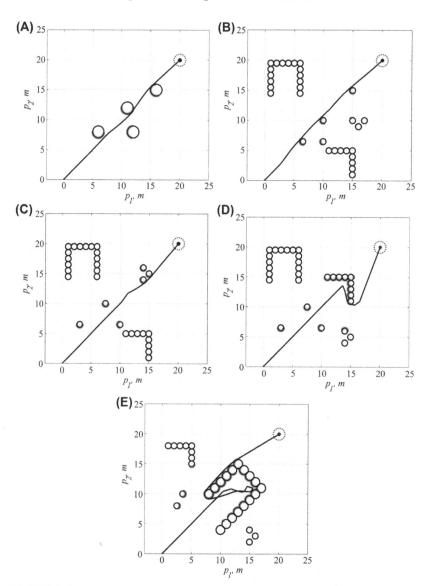

FIGURE 2.23 Results of the basic test-scenes passing using DVH-NN method: (A) scene 1; (B) scene 2; (C) scene 3; (D) scene 4; (E) scene 5.

as a solution. And for the case of several equidistant resulting values, the one with a maximal weight (index) will be selected. In other words, in the latter case the "right hand" rule will be obeyed.

It should be noted that NUCAC with a hardware priority encoder does not have the mentioned negative quality. Fig. 2.23 presents the modeling results performed in MatLab and demonstrating vehicle's successful motion in the test scenes using DVH-NN planning method. The experiments results presented

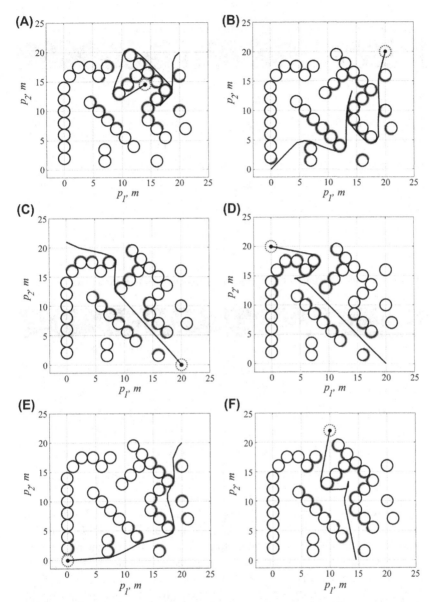

FIGURE 2.24 Results of scene 6 passing using DVH-NN method: (A) path from (20.0; 20.0) to (14.0; 14.5); (B) path from (0.0; 0.0) to (20.0; 20.0); (C) path from (0.0; 21.0) to (20.0; 0.0); (D) path from (20.0; 0.0) to (0.0; 20.0); (E) path from (20.0; 20.0) to (0.0; 0.0); (F) path from (14.5; 0.0) to (10.0; 22.0).

in Fig. 2.24 confirm the method's efficiency for a robot moving in a maze-type scene 6 for various combinations of initial conditions.

Values of statistical data corresponding to Figs. 2.23 and 2.24 are presented in Table 2.3. Similar to the previous example (see Table 2.2) the experiment statistics included the following data: scene type, set locator range, initial and goal

Table 2.2 DVH-Method Application Results for the Scenes 1—6

Parameter	Empty	Scene Number					
		1	2	3	4	5	6
Safety, m	—	0.05	0.12	0.05	0.05	0.04	0.05
Path length, m	28.27	28.69	28.35	28.85	34.21	45.00	34.83
Total controller time, s	8.90	12.80	12.50	11.70	14.40	22.40	16.90
Total search time Ts, s	8.92	12.83	12.53	11.73	14.44	22.45	16.94
Number of iterations	32.00	41.00	39.00	36.00	52.00	76.00	53.00
Mission success	1.00	1.00	1.00	1.00	1.00	1.00	1.00

Vehicle motion direction, from (0.0; 0.0) to (20.0; 20.0).

Table 2.3 DVH-NN Method Functioning Results for Scenes 1—6 for Various Initial Conditions of Vehicle's Motion Start

Parameter	Empty	Scene Number					
		1	2	3	4	5	6
Safety, m	—	0.07	0.12	0.05	0.05	0.05	0.04
Path length, m	28.27	28.36	28.35	28.42	34.26	46.31	52.63
Total controller time, s	8.90	13.30	12.50	11.50	14.40	25.20	32.10
Total search time Ts, s	8.92	13.33	12.53	11.53	14.44	25.27	32.18
Number of iterations	32	44	39	38	52	101	111
Mission success	1.00	1.00	1.00	1.00	1.00	1.00	1.00

point coordinates, minimal distance to the closest obstacle, passed path length, length of the shortest path to the goal with no obstacles, ISPPC controller functioning time, planner functioning time for a specific computer configuration, total task completion time, mission success estimate, and number of performed iteration cycles.

As in the case of DVH method, DVH-NN path finding algorithm can also get into a local minimum.

The DVH-NN method considered above allows fast creation of new coordinates in the locator active zone and of the vehicle's motion direction angle accounting for the current goal location. The method is based on the formal-logical basis of the classical neural-like networks with a homogeneous structure.

Method advantages: algorithm simplicity, high processing speed, and possibility of effective hardware implementation. The drawbacks are the same as for the DVH method considered in Section 2.3.2.

2.4 BIONIC METHOD OF NEURAL-NETWORKING PATH SEARCH

2.4.1 Planning Task Formalization for the Considered Method

Depending on the environment type and the used TVS, there can be various types of obtained information. Out of which two main classes can be separated. The first includes long range TVS forming only partial information about the situation in the "line of sight" zone. The second includes TVS with a space view providing information about the entire situation in the system activity zone. The TVS system of the second type produces data (space-time distribution of the reflected signal) that allows not only obstacle detection and distance measuring but also gives information about the situation behind the obstacles.

Grid representation of the environment model. The DVH and DVH-NN methods previously considered are intended for multibeam TVS usage. The bionic method of neural networking search presented below uses a TVS with a space sector view in the horizontal plane. The information is received periodically and transformed into a 2D discrete form with a preset accuracy, and then it is used for the vehicle path planning during motion inside the TVS active functioning zone.

The transformation into a 2D discrete representation is performed by preliminary binding a polar coordinate system to the sector view vertex followed by a step-by-step discretization of the entire sector zone by angle and distance. It should be noted that discretization step for the sector zone depends on the resolution and range of the TVS, clearness of the received signal, vehicle's speed, computing power of its ISPPC, and built-in memory size, etc.

The examples of creation of a simplified model of the environment using the long range TVS and one with space view are presented in Figs. 2.25 and 2.26, respectively.

Each node element of the formed grid model of the environment can be associated with elemental areas ΔS (see Fig. 2.27) obtained as a result of 2D discretization of the active zone data obtained from TVS. Depending on the set detection threshold and calculated average intensity, density of the reflected signal of the elemental area can be defined as occupied by obstacle (ΔS_o) or as free ΔS_f. In this model each occupied area is denoted with a black graph node and each free one with a hollow (white) one (see Fig. 2.27).

Vehicle motion planning task statement using bionic method and its formalization. To formalize the vehicle motion planning task in a Cartesian coordinated system $O_1P_1P_2$ let us introduce the following notation: X_{F_i} is the vector pointing toward the environment points that are free for vehicles motion; X_Π, vector of stationary obstacles points; X_{Π_i}, vector corresponding to the points of a mobile obstacle $\Pi(t_i)$ at the moment of time t_i; X_G, vector of goal position; X_{V_i}, vector of vehicle's position in the $O_1P_1P_2$ system at the moment of time t_i. The symbols T_1 and T_2 denote two robot's motion paths out of a number of possible ones.

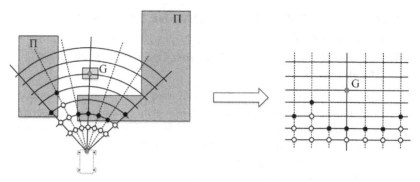

FIGURE 2.25 Simplified environment model building for a long range TVS.

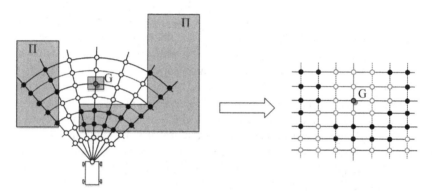

FIGURE 2.26 Environment model building for a space view TVS.

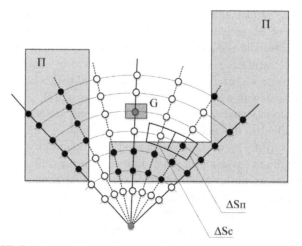

FIGURE 2.27 Selection of elemental areas in the sector-type locator's active zone.

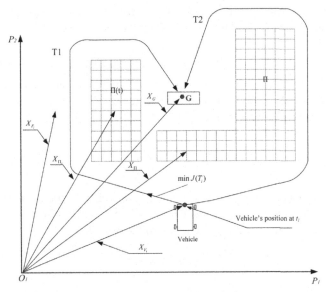

FIGURE 2.28 Part of the environment perceived by vehicle.

Fig. 2.28 shows the following denotations.
1. $t_i = t_{i-1} + \Delta t_i$ is the discrete time for environment analysis.
2. $X_{V_i}, X_{G_i}, \{X_{F_i}\}, \{X_{\Pi_i}\}$ are the parameters characterizing environment.
3. $L(X_i)$ is a function determining complexity of passing the environment sections by the vehicle. If the section is occupied by an obstacle, $L(X_i)$ takes the value of infinity. If the section is free $L(X_i)$ becomes equal to 0.
4. $W_i = W[X_{V_i}, L(X_i), X_G]$ are the modes of the environment passability, i.e., such model that isolates only one quality of the environment—the quality of being passable with a certain degree of difficulty or be totally impassable at the moment of time t_i.
5. $T_j(W_i) = T[X_{V_i}, L_j(X_i), X_{G_i}]$ is the jth path going from X_{V_i} to X_{G_i} accounting for the values of $L_j(X_i)$.
6. $|T_j(W_i)|$ is the norm of the path j, determining its complexity level for the robot.
7. To estimate the norm $|T_j(W_i)|$ the numbers $J(T_{ji})$ are introduced by the following relation:

$$J(T_{ji}) = \int_{X_{pi}}^{X_G} L_j(X_{F_i})dT_{ji}.$$

8. Any trajectory $T_j(W_i)$ is chosen out of a set of acceptable paths $T(W_i)$ at the moment of time t_i being an element of the set E_{Ti}, i.e., $T_j(W_i) \in E_{Ti}$. On the other hand, $J(T_{ji})$ is the number being an element of the set E_{Ji} of norms of all the acceptable trajectories, i.e. $J(T_{ji}) \in E_{Ji}$. Taking into account that

number $J(T_{ji}) \in E_{Ji}$ corresponds to trajectory $T_{ji} \in E_T$, mapping of the set E_{Ti} into the set E_{Ji} is a functional of the following type:

$$J(T_{ji}) = F(T_j(W_i)).$$

Then with all the introduced definitions, the motion planner's task is to take all the acceptable paths $T_j(W_i)$ at time moment t_i and find an optimal one $T^{opt}(W_i)$, satisfying the relation:

$$J\left(T_i^{opt}\right) = \min_j J(T_{ji}).$$

This task can be solved if at the point X_{V_i} we find a direction of antigradient of the functional $J(T_i)$, i.e., $-grad\ J(T_i)$, and determine the direction of the robot's speed V_i at the point X_{V_i} according to the relation:

$$V_i = -\rho\ grad\ J(T_i),$$

where ρ is the unity speed.

Note that vector $grad\ J(T_i)$ indicates the direction of maximum rate of increase of the functional $J(T_i)$, and vector $-grad\ J(T_i)$ indicates the maximal decrease rate direction at the initial point of the field of paths.

2.4.2 Development of the Method and of Neural-like Structures Implementing It

2.4.2.1 DEFINITION OF A FORMAL NEURON AS A THRESHOLD ELEMENT

Unlike the previous method, the FN based on the binary information representation will be used here as a main cell of the designed NN [12]. The concept of FNs is based on the notion of a neural cell being a logical element working according to the principle "all or nothing," i.e., it is either silent or gives out a similar signal. It is assumed that the cells have only axodendritic relations. Input and output spikes are approximated with unity impulses or unity potentials [12]. It is assumed that the output function of a neuron $y(t)$ is a Boolean function of Boolean variables $x(t)$, of the synaptic weights γj and of the threshold θ taking integer values [12,13].

Usually such an FN is defined as a *threshold logical element* with the following qualities:

1. it has N synaptic inputs that can be exciting ($\gamma j > 0$) or inhibiting ($\gamma j \leq 0$) with $j = 1, 2,..., N$;
2. the element states are researched at the equidistant moments of discrete time $t_i = t_0 + i \cdot \Delta t$, $(i = 1, 2,...)$, where $\Delta t = t_i - t_{i-1}$ is the sampling step of the discrete time t_i;
3. regardless of the number of unity signals coming to the inputs of FN at the moment t_i, the element delays the output signal for one step Δt of discrete time t_i;

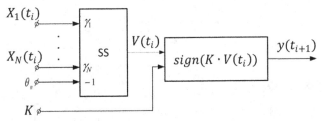

FIGURE 2.29 Block diagram of a formal-logical neuron.

4. each of the inputs $x_j(t_i)$ and outputs $y(t_{i+1})$ can be either in excited ($x_j(t_i) = 1$, $y(t_{i+1}) = 1$) or unexcited ($x_j(t_i) = 0$, $y(t_{i+1}) = 0$) state;
5. an FN has an excitation threshold θ. At the moment t_i if an algebraic sum of the exciting and inhibiting input actions is grater or equal than θ, $y(t_{i+1}) = 1$, otherwise $y(t_{i+1}) = 0$.

Thus, logic of the FN can be described as follows:

$$y(t_{i+1}) = sign(K \cdot V(t_i)), \tag{2.5}$$

where $V(t_i) = \sum_{j=1}^{N}(\gamma j \cdot Xj(t_i)) - \theta$; K is the parameter characterizing excitability of the neuron, and the function $y(t_{i+1})$ is a signum function taking values out of the set $\{0, 1\}$ (Eq. 2.6). If in (Eq. 2.5) we assume $\gamma j \in \{0,1\}$, the block diagram of an FN will take the form shown in Fig. 2.29.

$$y(t_{i+1}) = \begin{cases} 1, \text{if}(K \cdot V(t_i) > 0) \\ 0, \text{if}(K \cdot V(t_i) \leq 0) \end{cases} \tag{2.6}$$

It is rather simple and consists only of two units: unit of a space summation (SS) and a unit of output function $sign\ KV(t_i)$ that have an impulse or static type.

In the first case, if the excitation condition is satisfied ($K \cdot V(t_i) > 0$), the neuron's output signal is a unity of a priori set length τ. After the end of this signal, FN goes to an unexcited state and remains in it until the next satisfaction of the excitation condition. In the second case, if the condition ($K \cdot V(t_i) > 0$) is satisfied, the neuron goes to a unity state and stays there until the excitation condition is violated.

2.4.2.1.1 Development of a Neural Networking System of Afferent Synthesis

As discussed earlier, in the simplest case, when the environment is a surface S it can be split into elemental areas ΔS_j (see Fig. 2.27) and then transformed into an orthogonal grid model having a form of a flat orthogonal graph $G(Q,F)$, where Q is a set of nodes q_j corresponding to discrete areas ΔS_j, and F is the multitude of graph edges connecting these nodes (e.g., a graph with orthogonal links presented in Figs. 2.25 and 2.26). Each node q_j of the graph $G(Q,F)$ is associated with a neural-like processing element (PE). The separate elements are connected by informational links just like the nodes of the graph $G(Q,F)$ are connected by

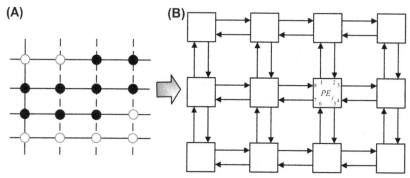

FIGURE 2.30 Topological structure example for a neural-like grid of a regular type: (A) graph structure $G(Q,F)$; (B) neural-like PE grid structure.

edges. As a result we obtain a neural-like network with its architecture presented in Fig. 2.30. Then each neural-like processing element, except for an element PE_0 corresponding to a node q_0, should perform operations as listed below:

1. If an element PE_j corresponds to a free area, it should conduct the excitation signals from the neighboring elements to the other ones connected to this PE.
2. If an element PE_j corresponds to an occupied area of the environment, it should block the signal distribution from the connected neighboring elements.
3. If an element PE_j corresponds to a goal area of the environment, it should generate excitation signals that can be distributed through other unblocked elements.

The processing element PE_0 is marked out because it corresponds to an environment area where the vehicle is located at any moment of time t_i. So PE_0 should perform the functions of a decision-making subsystem, i.e., it should be able to determine the direction of the first excitation signal and lock it for a period of time necessary for a vehicle effector subsystem to perform elementary motion in a required direction.

Assuming that input and output signals of the neural-like processing elements are Boolean variables, the function $y_j(t_i + \Delta t)$, implemented by the jth element can be written as follows:

$$y_j(t_{i+1}) = \overline{d} \wedge \left[\left(\bigvee_{\sigma \in I} y_\sigma(t_i) \right) \vee G_j(t_i) \right] \wedge \overline{\Pi}_j(t_i), \qquad (2.7)$$

where $y_\sigma(t_i)$ is the values of functions produced by the neighbor elements PE_σ directly connected to PE_j at the moment of time t_i; I, a set of indexes of such elements; $\Pi_j(t_i)$, binary signal incoming from environment model subsystem and describing the state of the corresponding area ΔS_j of the environment (if $\Pi_j(t_i) = 0$, it is free; if $\Pi_j(t_i) = 1$, it is occupied); $G_j(t_i)$, binary signal formed by the vehicle's ISPTC describing the goal presence at the area ΔS_j; d, binary signal coming to all elements of the neural-like network from the vehicle's ISPTC. On each cycle, the function $y_j(t_{i+1})$ is calculated. Before that the signal

d takes a banning value ($d = 1$) and resets the previously active elements. In the process of calculation of the function $y_j(t_{i+1})$ in the grid, the signal d has a permitting value ($d = 0$).

2.4.2.1.2 Development of the Decision-Making Unit

In order to make a decision about the robot's motion direction, it is necessary to use a special processing element PE_0 at the output of the NN that has n outputs and n inputs implementing the following logical expression:

$$Zj(t_{i+1}) = y_j(t_i) \left(\overset{n}{\underset{\substack{\sigma=1 \\ \sigma \neq j \\ \sigma}}{\wedge}} Z_\sigma(t_i) \right) \tag{2.8}$$

where $Zj(t_{i+1})$ ($j = 1,2,...,n$) is the signal at the jth output of PE_0; $y_j(t_i)$, signal at the jth input of PE_0.

It should be mentioned that processing elements implementing functions (2.7) can be represented by the considered neurons of formal-logical type (see Fig. 2.29) that have not only exciting inputs with a unity threshold but also two inhibiting inputs taking signals Π_j and d. The graphical representation of such an FN is shown in Figs. 2.31A.

Fig. 2.31B presents commutation of the FN inputs and outputs in its application as a processing element PE_j. Connecting PE_j according to the scheme in Fig. 2.30, we obtain a neural-like network modeling afferent synthesis. The outputs of the goals G_i, obstacles Π_i and signal d are not shown in Fig. 2.30.

It is interesting that functions of the special processing element PE_0 can also be implemented by a group of n FNs joined into an additional layer with a structure as shown in Fig. 2.32.

As presented in Fig. 2.32, each FN$_j$ ($j = 1, 2, ... , n$) is connected to the rest of the neurons of the layer by blocking the lateral links. This way the layer can implement the function (2.7) of the decision-making structure, i.e., it can detect and hold the excitation of the first fired neuron.

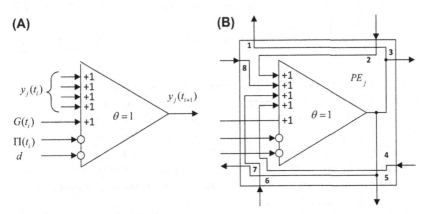

FIGURE 2.31 Example of the neural-like processing element (B) based on FN (A).

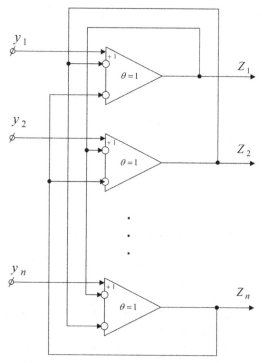

FIGURE 2.32 Formal representation of the neural layer structure based on FN, ensuring decision making ("The winner takes it all").

In the initial state, the neural-like network receives a blocking signal $d = 1$ at the inputs $y_i(t_i)$, and the outputs $Z_i(t_i)$ of the ensemble become zero. The inverted zero signals from the outputs $Z_i(t_i)$ go to the corresponding FN inputs which become excited if they receive a permissive signal d ($d = 0$). If $d = 0$, all FNs of the ensemble reach a state when any jth first unity input signal $y_i(t_i)$ would excite the output of FN_j; its output signal would in turn go to the blocking inputs of the rest of the FNs, thereby inhibiting their excitation. Therefore, other input signals $y_i(t_i)$ would no longer influence the work of the system until it is reinitialized with a signal d.

The algorithm of the neural networking decision-making layer is presented in Fig. 2.33.

2.4.2.1.3 Development of the Neural Networking Planner

Joining the FNs of the afferent synthesis system gives us a bionic neural networking planner (BNNP). At any moment of time t_i, it solves the task of determining the vehicle's motion direction in a non-formalized environment according to the approach described in Section 2.1 (see Fig. 2.2).

Information about the environment as a group of signals $G_j \in \{0,1\}$ and $\Pi_j \in \{0,1\}$ comes from the TVS to the afferent synthesis system with processing elements implementing function (2.7).

Processing elements of SAS are implemented using FN and are connected as shown in Fig. 2.30. The outputs of the *PE* of the last (top in Fig. 2.34) SAS

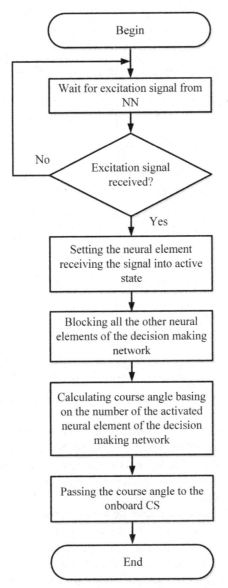

FIGURE 2.33 Extended algorithm of the decision-making layer.

row are connected not only to the other neurons of the structure but also to the excitation inputs (+1) of DMS FNs. The output signals of DMS move to the position-path controller ensuring a timely and appropriate adjustment of the parameter values of the implemented control law.

An example of a simplified BNNP is presented in Fig. 2.34. Let us describe its functioning.

Once robot's motion step ends, its effector system generates a short signal $d = 1$, deexciting all the FNs in SAS processing elements.

The processing elements of the unexcited SAS get the information about the obstacles location in the locator's active zone from TVS. For the preliminary

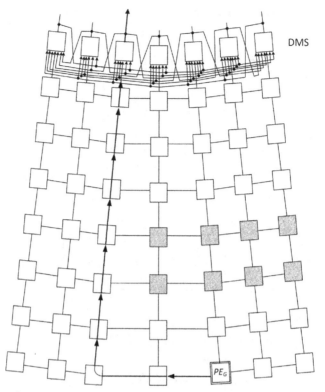

FIGURE 2.34 Simplified structure of bionic neural networking planner.

processing of the data from TVS and for the formation of the environment grid model, a system is installed at the input of the BNNP that implements the necessary actions of preprocessing and loading the model data into the PEs of SAS (in Fig. 2.34 the environment model formation subsystem (EMFS) is not shown). PEs that correspond to the occupied areas ΔSj of the environment are blocked, and the ones corresponding to the goal area is excited by the input signal $G = 1$ and becomes an excitation source for all the unblocked PE of SAS. The unity signals from its output are spread throughout the SAS only by conducting PEs constructing all the possible paths leading to the goal from DMS.

Obviously, the signal that takes the shortest path reaches the DMS first. DMS registers the direction of the last section of this way (angular deflection from the central axis of the locator's sector zone) and gives this information to the position-path controller. This will be the vehicle's motion direction in the next step.

As a result of robot's motion its position will change with respect to the obstacles and the goal. The information regarding the change will be perceived by the TVS, and then it will be processed and loaded into SAS after receiving the permissive d signal.

The processes described above are then repeated. So the ISPPC should include a unit (software–hardware module) performing real-time monitoring and forming the d signal of necessary level for SAS, ensuring synchronous functioning of the afferent synthesis system and the position-path controller.

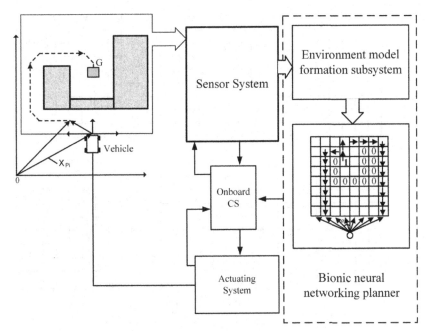

FIGURE 2.35 Application of BNNP for the vehicle's motion planning.

2.4.3 Method Modeling in MatLab

Using the theoretical foundations presented in the previous section, let us present the structure of the vehicle's motion planning system in a form of scheme as shown in Fig. 2.35.

In order to organize a software process of finding an optimal trajectory T_i^{opt} and a search for a unity vector of motion u_i^{bas} in the model, let us use an algorithm based on Lee wave algorithm for maze routing problems.

The essence BNNP modeling using the wave algorithm is described as follows:

1. In the goal array, a cell number with a goal entry is determined. Starting at this cell, a step is made upward, downward, right, and left.
2. Conjunction of the array elements of M_{W_i} and M_{Π_i}.
3. Each of the entries left in the array M_W after conjunction is spread again one step upward, downward, right, and left.
4. Conjunction of the arrays M_W and M_Π is performed again.
5. These processes are repeated until first entry gets to the lower cell C_0 of the array M_G associated with the location $X_{pi} = Y_0$ of the robot in the environment.
6. Number of unity bit in the cell C_0 determines the direction of the vector u_i^{opt}.
7. After the step ΔX_{pi} is completed in the direction of u_i^{opt}, the new content of the arrays $(M_{G_{(i+1)}}, M_{\Pi_{(i+1)}})$ is refreshed and a jump to step 1 is performed.

Let us illustrate functioning of this algorithm by the example presented in Fig. 2.36.

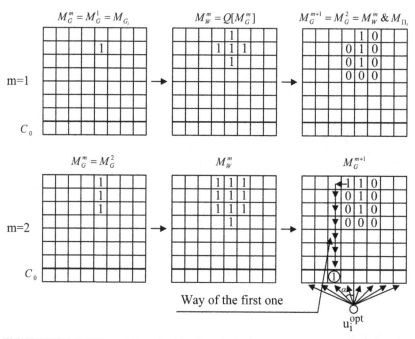

FIGURE 2.36 BNNP modeling algorithm functioning.

The array M_G^{m+1} holds the way of the first unity reaching the cell C_0. The angle α determines the direction of the unity vector u_i^{opt}.

The block diagram of the considered algorithm can be presented as shown in Fig. 2.37. The considered system allows to perform motion of the vehicle in environment with randomly moving obstacles and, in general, a moving goal.

Modeling results for this method performed in MatLab are discussed below.

As in the previous cases we will limit our experiments by six test scenes (scenes 1–6). For each scene we define two pairs of coordinates:

(p_1^0, p_2^0)—coordinates of the vehicle's starting motion point: (0.0; 0.0);

(p_1^k, p_2^k)—vehicle's goal coordinates: (20.0; 20.0).

As in the previous case 6 (see Fig. 2.9), we expand the values of the experiment's initial conditions:

the values (p_1^0, p_2^0) are selected from an ordered five-element set {(20.0; 20.0), (0.0; 0.0), (0.0; 21.0); (20.0; 0.0); (20.0; 20.0)};

the values (p_1^k, p_2^k) are selected from an ordered five-element set {(14.0; 14.5); (20.0; 20.0); (20.0; 0.0); (0.0; 20.0); (0.0; 0.0)}.

Let us set the limitations on the linear course speed (m/s): the maximum allowed value $V_{max} = 1.0$, permitted if there are no obstacles in the lidar visibility range. As the vehicle approaches its goal, the speed is gradually reduced in relation to the remaining distance. If the obstacles are detected, the speed is reduced by a factor of 5. As a result of preliminary experiments for passing the scenes successfully, the speed was reduced by a factor of 2.

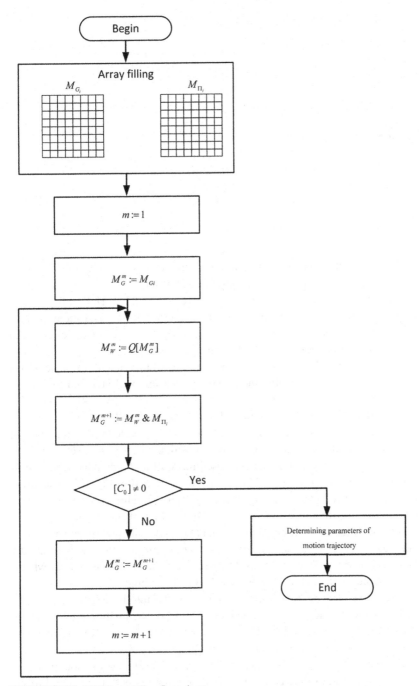

FIGURE 2.37 BNNP modeling flow chart.

```
if (IsSafetySituation(lidar_arr, 5, 45))
            if dzxy < deltaR
                % reduce speed ~1/2 one half of remaining distance
                    V = dzxy*(Vmax/deltaR)/2;
            else
                    V = Vmax/2;
            end
    else
            if dzxy < deltaR
                % reduce speed ~1/2 one half of remaining distance
                    V = dzxy*(Vmax/deltaR)/2;
            else
                % reduce speed to 1/5 of maximum value
                    V = Vmax/5;
            end
    end
```

Let us set the analogous limitations on the angular velocity of the "tank" design of the robot (deg/s): maximum allowed value $\omega^{max} = 25.0$.

Radius r of each of the robot's wheels is equal to 0.2 m. Half of the chassis length is equal to 0.5 m; the lumped mass center is located over the middle of the robot's chassis. This means that if $\{|\omega_l|; |\omega_r|\} \leq \omega^{max}$ in 1 s the chassis can rotate by not more than $\vartheta_\alpha = 10$ degrees.

Let us assume that locator (lidar) range is equal to 5 m and the value of δ is 0.02. So the values exceeding the maximum possible will be equal to 5.02 m. The other lidar parameters $\Omega = 91$ degrees, $\lambda = 1$ degree, $N = 91$.

Let us set the angular widening of the obstacles $\eta = 3$ degrees.

We set the values of Δp_1, Δp_2 as 0.02 m. As the vehicle moves from current coordinates to the predicted ones, it has a positioning error of ξ_r set to 0.02 m.

An object-oriented approach was used for MatLab modeling of the neural-networking planning method. The algorithm presented in Fig. 2.37 was used. Calculation of the predicted coordinate (p_1^{i+1}, p_2^{i+1}) values was performed by the classical procedure using sine and cosine values of the course angle φ_{i+1} generated by the NN. Based on this data, the vehicle path parameters are calculated using Eqs. (1.9) and (1.10). These parameters together with kinematic Eqs. (1.3) and (1.13) are then used to determine the speeds of coaxial wheels of the vehicles kinematic scheme of vehicle's modeling.

Fig. 2.38 presents the display windows of the interim experiment results such as scene's general view and vehicle's path (a), locator active zone content (b), scene representation by TVS data (c), and the results of NN visualization with a recommended motion direction (d).

The experiments results presented in Fig. 2.39 prove the method effectiveness in application to the problem of finding a robot's path from initial to the final point in the scenes 1−5 for the same initial conditions.

The results of passing the complicated scene using the considered method for various initial conditions are presented in Fig. 2.40.

As in the previous cases, the obtained results are combined in Table 2.4. The same data as in the previous experiments is presented.

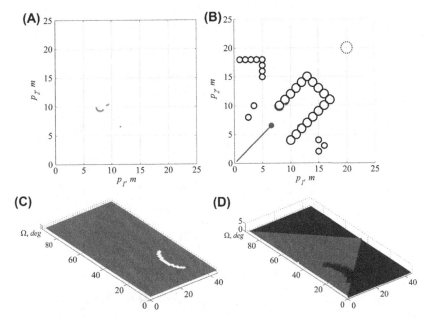

FIGURE 2.38 Algorithm functioning results: (A) scene general view and motion path; (B) active locator zone content; (C) TVS scene view; (D) neural network content.

Despite the fact that for passing a complicated maze (scene 5) and for looping avoidance the speed was reduced by a factor of 3 and the obstacles were widened up to 7 degrees, we could not avoid getting into a local minimum (see Fig. 2.40F). In case of maze-type scenes the situations can repeat causing looping of the vehicle's motion. To avoid such situations there must be an intelligent level recognizing the situation of local minimum and generating a new motion direction making it possible to avoid looping in future.

The main advantages of the method are its simplicity, low computational and hardware costs, possibility of paralleling, and easy integration with other algorithms.

The major drawbacks are:
- possibility of search procedure looping;
- law safety index and its direct relation to the value of obstacle's angular widening;
- dependence of the effectiveness of the solutions taken by the planner from the current coordinates of the vehicle and its goal during the robot's motion process;
- due to the peculiarities of the space—time discretization of the environment, the path can be planned only for a short distance from the locator resulting in a stepping character of the path at the distances far from the vehicle;
- strong dependence of the planning quality from the locator parameters (e.g., lidar) and from the current robot's motion speed.

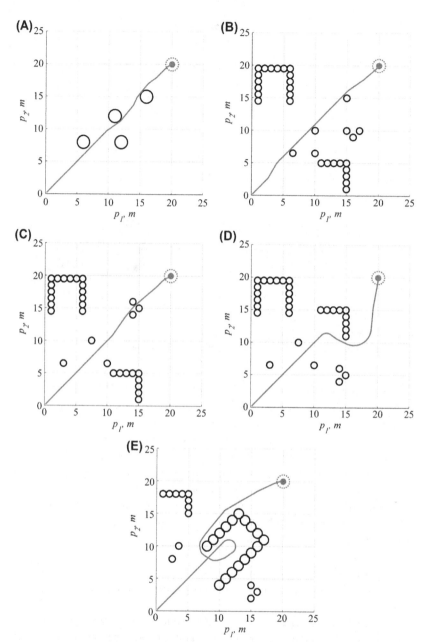

FIGURE 2.39 Scene passing results: scene passing examples: (A) scene 1; (B) scene 2; (C) scene 3; (D) scene 4; (E) scene 5.

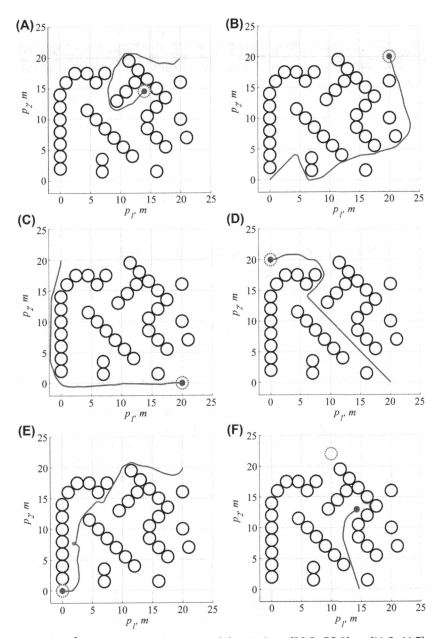

FIGURE 2.40 Scene 5 passing results: (A) path from (20.0; 20.0) to (14.0; 14.5); (B) path from (0.0; 0.0) to (20.0; 20.0); (C) path from (0.0; 21.0) to (20.0; 0.0); (D) path from (20.0; 0.0) to (0.0; 20.0); (E) path from (20.0; 20.0) to (0.0; 0.0); (F) path from (14.5; 0.0) to (10.0; 22.0).

Table 2.4 Neural-Networking Hybrid Method Functioning

Parameter	Empty	Scene Number					
		1	2	3	4	5	6
Safety, m	—	0.09	0.10	0.10	0.04	0.05	0.11
Path length, m	28.27	28.35	28.33	28.33	32.16	46.47	44.81
Total controller time, s	12.67	23.84	22.40	21.46	25.90	53.41	74.88
Total search time Ts, s	12.77	23.94	22.50	21.56	26.00	53.51	74.98
Number of iterations	83.00	147.00	139.00	134.00	139.00	225.00	247.00
Mission success	1.00	1.00	1.00	1.00	1.00	1.00	1.00

2.5 CONVOLUTIONAL NEURAL NETWORKS[1]

2.5.1 Path Planner Structure

This section presents an approach to planner development for robotic vehicles based on CNN.

To apply the stated problem solution and according to the research performed in works [14—16] the structure of the neural networking planner takes the following form (Fig. 2.41).

The input data for the planner is the image from the TVS (locator in our case) and goal position. The motion goal is a certain point in space where the motion is directed or a path that should be followed. The planner mode changes according to the motion goal. Since map is not used according to the planner requirements, there is no initial data concerning the environment. The data from the TVS contains the current information about the local functioning area. Using this information, the planner should correct the vehicle's motion path (angle and speed) to avoid any detected obstacles.

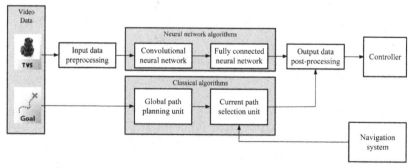

FIGURE 2.41 Functional scheme of neural-networking planner.

---

[1]The work of M.U. Sirotenko [100] was used in writing of this section.

The preprocessed data from TVS goes to the input of the CNN that forms the feature map that corresponds to the current situation. The goal information is fed to the global path construction unit (GPCU). Depending on the goal type this unit can function in two modes. In positioning task, GPCU uses the classical path search methods and forms a motion path as a mission scenario from the current point to the goal one defining various motion sections using quadratic and/or linear forms. When solving a path-following task, GPCU determines the shortest distance from the current position to the trajectory and forms an appropriate mission scenario [14]. In order to form a global trajectory the classical methods are used in most of the cases, such as wave algorithms of modified versions of A*.

In order to form a motion path based on TVS data, a CNN is proposed to be used [16]. A CNN is a special class of NN widely used for video and audio data treatment. CNNs combine three architectural ideas to ensure invariance to shift and distortion of the initial image: local receptive fields, shared weights, and spatial subsampling [17]. CNN consists of four alternating layers of convolution and subsampling. The image received by the CNN is convolved with a certain kernel according to the following expression:

$$C\{n\}(i,j) = F\left[\sum_{k=1}^{ks}\sum_{l=1}^{ls}K\{n\}(k,l)\cdot S(i-k,j-l)+B(n)\right]. \qquad (2.9)$$

where C is the convolution result (feature map); n, layer number; i, j, indexes determining location of an element in a feature map; F, saturation function that usually can be tangential or sigmoidal; k, l, indexes determining position of an element in convolution kernel; ks, ls, kernel size; K, convolution kernel; S, input image; and B, shifts matrix.

The convolution kernel is a set of shared weights. The result of such an operation is a certain image that is called a feature map. Depending on the kernel selected, the feature map can extract elementary visual features from the input image. Several different kernels are used for this purpose resulting in several feature maps.

The convolution layer is followed by a layer of averaging and subsampling that reduces the feature map's dimension reducing the output sensitivity to shifts and rotations.

Alteration of convolution and subsampling layers results in a gradually growing number of feature maps with their dimension reducing from layer to layer.

The outputs of the convolution layers are fed to the inputs of the classical multiply connected NN to form a motion path in a form of rotation angle and motion speed.

For an effective functioning of a neural networking planner, it is necessary to make a correct selection of the number of layers and neurons in the layers of convolutional and classical multiply connected NN. The layers number selection for the convolutional network is determined by the initial data dimension and complexity of the sought visual features (feature maps).

2.5.2 Methods of Visual Planning and Obstacle Avoidance

Visual planning is an extension of the considered methods of planning and obstacle avoidance for the tasks using a camera-based TVS [18] or a scanning system as a source of information about the environment.

Most of the works in this area are devoted to the research of such problems as optimal landmark selection for robot's localization [19], recognizing the current robot's position using visible landmarks ("kidnapped robot problem") [20], tracking visual references and position adjustment [21], and path planning accounting for robot's sensory capabilities [22]. The works [23,24] address a problem of selecting the best robot's position giving the most adequate environment representation.

A separate class of tasks is related to the avoidance of visual obstacle. Generally, the systems implementing obstacle avoidance using visual information are called reactive. One of the first vehicles that had a camera-based system of obstacle avoidance was "Shakey the Robot" [25] built in Stanford University during the period from 1966 to 1972.

The avoidance system of this robot detected visual contours and made it possible for him to move in the environment with static obstacles contrasting with the floor. Similar approaches were used in Ref. [26] for navigation of robots Polly and Frankie created in MIT. The work [27] presents an approach that uses image's color information instead of contours. In Ref. [28] a method is proposed to detect the obstacles using a reference area (reference window) directly in front of the robot that is compared to the rest of the image.

In the work [29] it is proposed to use three independent units of obstacle detection using brightness gradient criterion, RGB and HSV colors. The results were combined to obtain obstacle edges positions and to generate control actions for the actuators. This approach requires color images at the system's input and sharp detection of the obstacle contours.

The work [30] presents a method of visual planning that proposes to transform the visual features into a configurational space where planning is performed.

A special emphasis should be made on methods using optical flows for visual recognition and obstacle avoidance [31–34]. An optical flow between a pair of images is a vector field defining a natural (in the widest sense) transformation of the first image into the second one. One of the most demonstrative in this class is the work [32]. The algorithm presented in it is based on the real-time analysis of foveal and peripheral vision, calculation of image signal derivatives with respect to spatial and temporal coordinates, and generation of control actions on this basis. A disadvantage of the proposed approach is the presence of a blind spot at the image center caused by zero derivative of stationary obstacles located on the vehicle's line of motion. In addition, in the real conditions correct functioning of the optical flow—based algorithms require mechanical compensation of the camera's vibrations. The absence of texture leads to low accuracy in calculation of optical flows [14].

2.5.3 Local Motions Planner

This section proposes a method of building a local motions planner that uses information of the TVS based on locator (lidar). The planner outputs provide the necessary parameters of the current motion path (see Section 1.5). It is proposed to use CNN that were trained in two stages.

The task of local planning based on visual information requires complex processing of hardly formalizable data in order to convert it into obstacle avoidance path parameters. This task combines two problems: a problem of scene classification and a problem of nonlinear function approximation transforming the image space into a space of angular and speed parameters with a significantly lower dimension.

The proposed approach for creation of neural networking local motions planner does not solve the problems of objects or scene recognition (or their correlation). It solves a problem of reflectory reaction on changing of the environment.

A well-known Hubel and Wiesel visual system research allowed to get a better understanding of visual cortex structure and encouraged using this knowledge in NN. The main concepts included the locality of receptive fields and separation of neurons according to their functions inside one layer.

Localization of perception means that the neuron receiving the information monitors only a part of the signal and not the whole range. Such a region of sensory space is called a neuron's receptive field.

The notion of a receptive field must be specifically clarified. Traditionally, a receptive field of a neuron is a set of receptors that affects the neuron's functioning. The receptors here are understood as neurons directly receiving the external signals. Imagine a NN consisting of two layers. The first one is the receptor layer and the second one is a set of neurons connected to receptors. For each neuron of the second layer, the receptors connected to it constitute its receptive field.

Now let us consider a complex multilayered network. As we move away from the input, it becomes harder to point out which receptors influence the deeper neurons and how do they do it. At a certain point it may happen that for a certain neuron all the existing receptors can be called its receptive field. In such a situation only the neurons being in direct synaptic contact are called its receptive field. In order to separate these notions, the input set of receptors is called an initial receptive field. The neurons in direct contact with the neuron are called its local receptive field or just a receptive field (without any clarification).

Separation of neurons according to their functions is connected to discovery of two main types of neurons in the primary visual cortex. Simple cells respond to stimuli located at the specific place of the initial receptive field. The complex cells respond to the stimuli regardless of the exact location.

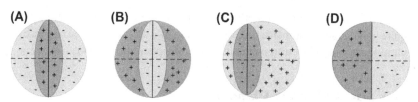

FIGURE 2.42 Initial receptive fields of a simple cell [46].

Fig. 2.42 presents the variants of sensitivity maps of the initial receptive fields of simple cells. The positive areas activate a neuron, the negative ones inhibit it. Each simple cell has a stimulus that suits it the best and causes the maximum excitation. But it is important that this stimulus is strictly tied to a position at the initial receptive field. A simple cell would not respond to the same stimulus shifted aside.

Complex cells also have a preferred stimulus but they would recognize it regardless of its position on the receptive field. These two ideas gave birth to the corresponding NN models. One of such networks was created by Kunihiko Fukushima and was called "cognitron." Later he created a more advanced network "neocognitron." Neocognitron is a multilayered structure with each layer containing simple (S) and complex (C) neurons.

The task of a simple cell is to monitor its receptive field and to recognize the trained pattern. The simple cells are united in groups (planes). Inside one group, the simple neurons are tuned to the same stimulus but each neuron monitors its own fragment of the receptive field. Together they move through all the possible positions of the pattern (see Fig. 2.43). All the simple neurons of one plane have the identical weights but different receptive fields. The situation can also be seen from a different point of view, we can envision a single neuron that can try its pattern on all the positions of the initial picture. All of this allows for recognition of the same pattern independent of its position.

Each of the complex cells monitors its plane of simple neurons and fires if any of the simple cells of its plane becomes active (Fig. 2.44). Activation of a simple cell means that it recognized a characteristic stimulus at the specific position of its receptive field, while activation of a complex neuron means that this pattern is found, generally, in a layer being monitored by simple cells.

After the input layer, each new layer inputs a picture created by the complex cells of the previous layer. The information is generalized from layer to layer, which results in recognition of specific patterns regardless of their location on the initial picture; and a recognition that is invariant to some transformation.

In 1995, as an extension of this idea, an American scientist of French origin, Yann LeCun proposed a separate type of NN—CNN [17] as a special class of NN that ideally matches the needs of intellectual processing of visual and audio data [35,36]. CNNs combine three architectural ideas to ensure invariance to shift and distortion of the initial image: local receptive fields, shared weights, and spatial subsampling.

FIGURE 2.43 Receptive fields of simple cells trained to search for a selected pattern in different positions [47].

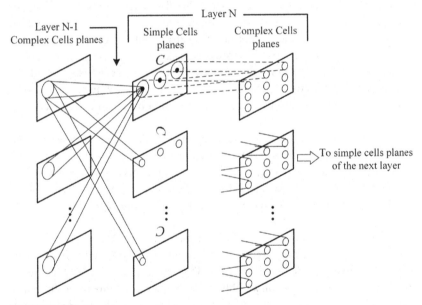

FIGURE 2.44 Neocognitron planes.

Each neuron of a layer receives data from a limited image section located in its monitoring area. With local receptive fields, neurons can extract elementary visual features such as oriented edges, end points, or corners. These features are united in maps for analysis by the higher layers. For visual features to be equally well recognized in different parts of the image, additional constraints on the weights are applied. Different sets of weights applied to the image form different feature maps that are combined and analyzed by the next levels yielding new feature maps. This operation is equal to convolution with a certain kernel. Each convolutional C-layer is followed by a subsampling S-layer that reduces the resolution of the feature map by a factor of 2 by averaging and applies a certain activation function to the result. Such alternation of layers increases the number of feature maps, while decreasing their spatial resolution.

The local planning problem based on the laser range finder data is nontrivial. This statement moves implicit scene analysis into first place. On one hand, this requires input image identification, while on the other hand the output should provide parameter values necessary to form the desired path $\left(\text{motion coordinates, course deviation angle } \varphi, \text{ linear course speed } V_k, \right.$ and maximum allowed values of the robot wheel angular velocities $\left(\omega_L^{max}, \omega_R^{max}\right)\right)$. The first task is solved by building a classifier and the second one by means of regression.

These tasks are solved together by two-stage training of the NN. At the first stage a NN with one fully connected output layer is formed. The inputs receive environment images corresponding to various classes of situations and the network is trained to classify the images. The first stage forms the weights of convolution layers sensitive only to invariant features of the given environment.

At the second stage the output layer is substituted by two fully connected layers. This allows us to solve the regression task in order to form the necessary path parameters (see Section 1.5) based on the invariant features obtained.

A sketch of a CNN is presented in Fig. 2.45. According to Ref. [14] a trained system with a "teacher" can be represented by the structure shown in Fig. 2.46.

The trained system calculates the function $M(Z,W)$, where Z is a set of input vectors of features, W is a set of trained parameters. The objective function E $(D, M(Z, W))$ (where $E = \{E_1, E_2, ...E_p\}$—a set of errors) estimates the difference between the current output values of the trained system M (where $M = \{M_1, M_2, ...M_p\}$) and the desired output values D (where $D = \{D_1, D_2, ...D_p\}$) respectively.

The CNN consists of two alternating types of layers: C-layer and S-layer. Let I be an input image having w points horizontally and h vertically. C-layer is set by its matrices of shared weights W_m^k and shifts B_m^k, where k is the C-layer's number, m, number of weights or shifts matrix for this layer. Introduction of several matrices of shared weights for one layer is done as it is necessary to search for different features in the same image. The size of matrix W can vary but usually is selected to be square with dimensions of 7×7, 5×5 or 3×3. The output of a convolution layer is a feature map FM. Forward propagation in the convolution layer is described by the following expression:

$$X_{m,i,j}^k = \sum_{q=1}^{Q} \left(\sum_{l=1}^{W_w} \sum_{n=1}^{W_h} W_{m,l,n}^k \cdot X_{q,i-l,j-n}^{k-1} + B_m^k \right) \qquad (2.10)$$

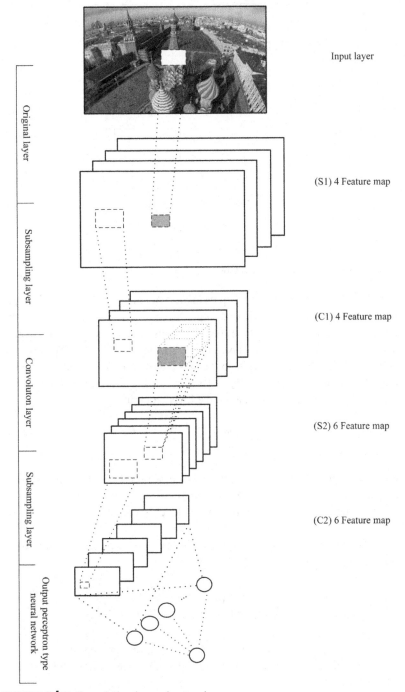

FIGURE 2.45 Convolutional neural network.

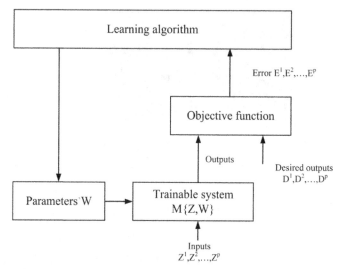

FIGURE 2.46 Trained system structure.

where X_m^k is the mth feature map (outputs matrix) of kth layer; Q, number of input feature maps of kth layer (number of output feature maps is $k - 1$); W_w and W_h, number of columns and rows of the shared weights matrix; and X_q^{k-1}, qth feature map at the output of the previous layer, for the first layer $X_q^{k-1} = I_q$. The result of operation (Eq. 2.10) is a feature map that has a dimension lower than the initial one by $W_w - 1$ horizontally and $W_h - 1$ vertically. The expression (2.10) can be rewritten for the entire feature map at the layer's output:

$$X_m^k = \sum_{q=1}^{Q} \left(conv\left(X_q^{k-1}, W_m^k\right) + B_m^k \right) \tag{2.11}$$

where $conv$ is the 2D convolution function. As we see from the expression, in such presentation the feature map at the output of the previous layer serves as a signal, and the weights matrix of the current level is a convolution kernel or 2D filter.

For the subsampling layer the process of forward propagation is described by the following expression:

$$X_{m,i,j}^k = F\left(\sum_{l=1}^{Sr} \sum_{n=1}^{Sr} W_m^k \cdot \frac{X_{m,Sr\cdot i+l,Sr\cdot j\cdot+n}^{k-1}}{Sr^2} + B_m^k \right) \tag{2.12}$$

where F is an activation function and Sr, subsampling degree. As a result of passing through S-layer the feature map becomes Sr times smaller horizontally and vertically.

When we use a NN applying it to different problems, one of the main tasks is to select its structure. At present, there is no universal method of selecting a number of layers and neurons in the layer for different tasks. The situation is aggravated by the fact that for the classical multilayered NN an excessive number of neurons and layers can lead to the effect of overtraining that negatively affects the process of approximation or classification. This led to the usage of network reduction methods such as optimal brain damage [37].

Owing to shared weights, CNN are practically immune to overtraining. That is why an extra number of layers and weights cause a situation when different sets of weights tune on the search for the same features and duplicate each other. The main negative effect of this is the planner's performance reduction complicating the real-time functioning. Thus, when we build a CNN there is information about the input image dimension and the requirements put on the output data. It is necessary to define the following parameters: number of layers; number of feature maps at the output of each layer; dimensions of convolution kernels for each C-layer; subsampling degree for each S-layer; and number of multiply connected layers and the number of neurons in them.

The network architecture is selected based on the type and dimension of the input image. It can be either gray scale or colored. The first type image is encoded by a 2D array of intensities having the size $W \times H$, where W is the image width (number of array columns) and H, image height (number of array rows). A colored image is encoded by a 3D array that has red, green, and blue components.

The CNN structure is a pyramid. From layer to layer, the feature map dimension is reduced with an increase in their number. This is performed in such a way that the dimension of feature maps at the output of the last convolution layer is 1×1. In such approach, the number of layers depends on the selected dimensions of convolution kernels. Feature map dimension reduction is a result of convolution and subsampling. Analyzing the literature on CNN [38–40] we can conclude that using kernel dimensions higher than 5×5 rarely gives a positive effect. Selection of feature maps quantity for each layer depends on the complexity of features detected by the network and is performed experimentally.

The planning problem is going to be solved using the architecture presented in Fig. 2.47. As we see from its structure, the spatial dimensions of the feature maps are reduced from layer to layer and their number increases. The output of this CNN is a vector consisting of 480 elements carrying the information about the space in front of the vehicle.

The output of the CNN is fed to the input of a fully connected NN with two sections having 100 and 2 hidden neurons and having 2 outputs forming the parameters of angle and speed that will be used to figure out the parameters of the vehicle's path.

As an activation function in S-layers, it is proposed to use a tangential function with input coefficient of 2/3 and output coefficient 1.7159, i.e.,

$$f(x) = 1,7159 \tanh\left(\frac{2}{3}x\right) \tag{2.13}$$

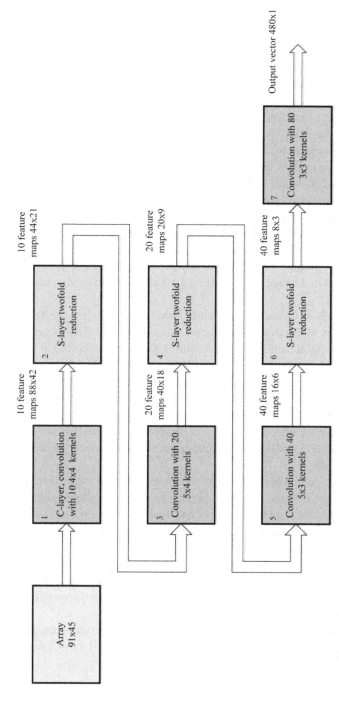

FIGURE 2.47 Architecture of a CNN for a local path planner.

Selection of a function with such coefficients gives the unity input gain equal to 1 [41], which is rather convenient for monitoring the system's dynamics.

A root-mean-square deviation (RMSD) function is going to be used as an objective function:

$$E^p = \frac{1}{2}(D^p - M(Z^p, W))^2 \qquad (2.14)$$

At present there are many ways of NN training. The two main methods include gradient and heuristic. The foundation of gradient methods is the Tailor expansion that, in case of RMSD (Eq. 2.14) function, has the following form [42].

$$E(W + p) = E(W) + [g(W)]^T p + \frac{1}{2} p^T H(W) p + \dots \qquad (2.15)$$

where p is the vector giving the direction of weights shifting; $g(W) = \nabla E = \left[\frac{\partial E}{\partial W_1}, \frac{\partial E}{\partial W_2}, \dots, \frac{\partial E}{\partial W_n}\right]$ is the gradient vector; and the matrix

$$H(W) = \begin{bmatrix} \dfrac{\partial^2 E}{\partial w_1 \partial w_1} & \cdots & \dfrac{\partial^2 E}{\partial w_1 \partial w_n} \\ \vdots & \ddots & \vdots \\ \dfrac{\partial^2 E}{\partial w_n \partial w_1} & \cdots & \dfrac{\partial^2 E}{\partial w_n \partial w_n} \end{bmatrix} \qquad (2.16)$$

is a Hessian matrix.

Zeroing a particular term in expression (2.15) we can influence speed and convergence of the training process. Neglecting all the terms except the gradient yields a fastest descent algorithm [43]:

$$p = -g(W) \qquad (2.17)$$

This method allows us to minimize the objective function (2.14). However, there is a possibility of getting into a local minimum and convergence speed can be unsatisfactory. The methods incorporating the Hessian matrix in calculation of p are called Newtonian:

$$p = -[H(W)]^{-1} g(W) \qquad (2.18)$$

Such methods account for the error surface curvature speeding up the convergence sufficiently. The probability of getting into a local minimum is reduced as well. Ideally, when the error surface is quadratic, the Newtonian method gives a solution that leads to a global minimum in one step. If the Hessian matrix is not positively definite, the Newtonian algorithm will diverge.

In a general case for the NN we cannot guarantee that the Hessian matrix is positively definite. In addition, calculating the inverse of a Hessian matrix has the complexity of $O(N^3)$. For these reasons the Newtonian methods are rarely used in practice. Quasi-Newtonian methods are used more frequently. One of the most widely spread and tried-and-true method is the Levenberg–Marquardt algorithm. Its idea is the substitution of the Hessian matrix $H(W)$ by its approximation $G(W)$ obtained using the knowledge of the gradient. If the purpose function is written in the following form

$$E(W) = \frac{1}{2} \sum_{i=1}^{m} [e_i(W)]^2 \qquad (2.19)$$

where $e_i(W) = (X_i(W) - D_i)$, for kth training step the training vector will be equal to

$$p_k = -[G(W_k)]^{-1} g(W_k), \qquad (2.20)$$

$$G(W_k) = [J(W_k)]^T J(W_k) + \mu, \qquad (2.21)$$

where μ is the regularizational factor that decreases as the minimum is approached; $J(W_k)$, Jacobean matrix defined as follows:

$$J(W) = \begin{bmatrix} \dfrac{\partial e_1}{\partial w_1} & \cdots & \dfrac{\partial e_1}{\partial w_n} \\ \vdots & \ddots & \vdots \\ \dfrac{\partial e_M}{\partial w_1} & \cdots & \dfrac{\partial e_M}{\partial w_n} \end{bmatrix}. \qquad (2.22)$$

Approximation of the Hessian matrix allows substantial reduction of the computational resource requirements and ensures its positive definition necessary for convergence of the training process.

Selection of the algorithm for NNPP is not as obvious as for the classical fully connected sigmoidal NN. The reason is that the shared weights in the network structure make them weakly sensitive to retraining. So the best results can be achieved in training by excessive amounts of data consisting of thousands and tens of thousands of training pairs. This circumstance has several effects:

- calculation of the genuine gradient and genuine Hessian matrix is complicated;
- the process of forming the training sample group must be automated to the maximum; and
- training algorithm speed is very important because speedup by 1% can save many hours of calculations.

The first consequence is explained by the fact that calculation of the genuine gradient requires the entire training sample. Such a training is called batch training. An alternative approach is to use stochastic methods that calculate an estimate of the genuine gradient using a specific training sample. The samples are selected in a pseudo-random manner which reduces the chances of the stochastic gradient method getting into the local minimums. In addition, it works much faster than the batch training methods and can be used for online training. Another advantage of the stochastic gradient method is that it provides faster training for excessive data. On the other hand, a drawback of stochastic methods is the impossibility to use second order methods for convergence speedup.

For a neural networking planner, it is proposed to use a stochastic diagonal modification of the Levenberg–Marquardt algorithm [44]. This method has two differences allowing us to use it in a stochastic mode. The first one is in accounting for only diagonal elements of the Hessian matrix, i.e.,

$$\frac{\partial E_1}{\partial w_i^2} = \sum_k \frac{\partial^2 E}{\partial o_k^2} \left(\frac{\partial o_k}{\partial w_i} \right)^2 \tag{2.23}$$

where o_k is a kth output of the NN. The expression (2.23) allows us to calculate the values of the Hessian diagonal matrix for the output layer weights. In order to calculate the matrix of all the weights and layers, a back-propagation algorithm is used that is analogous to gradient back-propagation:

$$\frac{\partial^2 E}{\partial y_k^2} = \frac{\partial^2 E}{\partial o_k^2} (f'(y_k))^2 \tag{2.24}$$

$$\frac{\partial^2 E}{\partial w_{ki}^2} = \frac{\partial^2 E}{\partial y_k^2} x_i^2 \tag{2.25}$$

$$\frac{\partial^2 E}{\partial x_i^2} = \sum_k \frac{\partial^2 E}{\partial y_k^2} w_{ki}^2 \tag{2.26}$$

where y_k is the weighted sum of the kth neuron's inputs.

Since for the neural networking planner an RMSD (2.14) function is selected to be an objective function, and a tangential function as an activation function, substitution of Eq. (2.14) into Eqs. (2.24)–(2.26) for C-layer yields

$$\frac{\partial^2 E}{\partial y_k^2} = y_k^2 \tag{2.27}$$

$$\frac{\partial^2 E}{\partial w_{ki}^2} = y_k^2 x_i^2 \tag{2.28}$$

$$\frac{\partial^2 E}{\partial x_i^2} = \sum_k y_k^2 w_{ki}^2 \qquad (2.29)$$

From the expression (2.22) it follows that matrix $\frac{\partial^2 E}{\partial W^2}$ will include derivatives for all the links between the outputs of the previous x_i and the inputs of the current layer y_k. But this expression does not account for the fact that the weights are shared and their number is significantly smaller than the number of links. With this correction, the expression (2.22) is transformed to the following form:

$$\frac{\partial^2 E}{\partial w_{m,u,v}^2} = \sum_l \sum_{i,j} y_{i,j}^2 x_{i+u,\, j+v}^2, \qquad (2.30)$$

$$\frac{\partial^2 E}{\partial W_m^2} = \sum_l conv\left(rot180\left(X^2\right), Y^2\right), \qquad (2.31)$$

where $rot180$ denotes 180 degrees matrix rotation function, l, feature map number, and m, convolution kernel number. Analogously, the expression (2.29) is transformed into the following form:

$$\frac{\partial^2 E}{\partial X_l^2} = \sum_l conv f\left(W^2, Y^2\right) \qquad (2.32)$$

where $conv f$ is the so-called full 2D convolution that adds the convoluted matrix with $kw - 1$ columns and $kh - 1$ rows of zeroes, and then performs a standard 2D convolution. The proposed form of error back-propagation algorithm for a neural networking planner based on a CNN allows us to increase the training speed if modern computing facilities are being used.

The second feature of this method is the calculation of the current estimate of Hessian diagonal matrix. The initial value of the Hessian matrix is calculated according to the expressions (2.27)–(2.29). Later on, the learning coefficient for each weight can be calculated according to the following expression:

$$\eta_{ki} = \frac{\epsilon}{\left\langle \frac{\partial^2 E}{\partial w_{ki}^2} \right\rangle + \mu} \qquad (2.33)$$

where ϵ is the global learning coefficient; $\left\langle \frac{\partial^2 E}{\partial w_{ki}^2} \right\rangle$, current estimate of the diagonal of the errors second derivative with respect to weights; and μ, coefficient preventing sudden growth of the η_{ki} value if the second derivative is small, i.e., when optimization takes place in a smooth section of the error function. The current estimate of the second derivative is calculated according to the following expression:

$$\left\langle \frac{\partial^2 E}{\partial w_{ki}^2} \right\rangle_{new} = (1 - \gamma) \left\langle \frac{\partial^2 E}{\partial w_{ki}^2} \right\rangle_{old} + \gamma \frac{\partial^2 E}{\partial w_{ki}^2} \qquad (2.34)$$

where γ is the constant determining the changing speed of the current estimate. Such an approach results in a significant training speedup at rather low computational costs.

The initial weights of the NN have a rather significant effect on the training process. A tangential function with linear and saturation segments is used as an activation function. The weights are selected in a pseudo-random manner in such a way that would make the activation function work in its linear segment. If the weights are too large, the tangential function can get saturated causing low gradient values and very slow training. If the weights are too small, the gradients become small too. The average weight values, first make the gradients large enough to pass the training and, second make the network learn the linear transforms before proceeding to more complex nonlinear ones.

Assuming that input data of an NN are normalized, in order to ensure the described network functioning mode it is necessary for each neuron to have a distribution's standard deviation σ approximately equal to 1. Assuming that the neuron's inputs are not correlated, the standard deviation of the weighted sum of the neuron's inputs is given by the following expression:

$$\sigma_{y_t} = \left(\sum_j w_{ij}^2 \right)^{\frac{1}{2}}$$

Therefore, for the standard deviation to be approximately equal to 1, the weights have to be pseudo-randomly initialized with a zero expectation value and a standard deviation given by:

$$\sigma_{y_t} = m^{-\frac{1}{2}}$$

where m is the number of neuron's inputs.

2.5.4 Motion Planner Modeling

An approach based on using the virtual scenes modeling instruments is proposed to be used for creation of a teaching sample in the modeling mode.

At the first stage MatLab package is used for creation of a scene where the obstacles are located. The obstacle coordinates are changed pseudo-randomly inside the locator range generating a set of scenes with different mutual positions of the vehicle and obstacles. Based on these scenes, a data (image) flow from the locator is generated during the vehicle's motion toward the obstacle.

In order to ensure a correct training of the neural networking planner, a sample must be created ensuring maximum variation of all the elements of the scenes.

As an example, it is proposed to consider the task of training the planner to create paths avoiding mobile obstacles—objects appearing on the vehicle's line of motion.

For this purpose a special scene was created as presented in Fig. 2.48.

In order to form a training sample it is necessary to have images from the TVS. Besides, the path coefficients corresponding to these images have to be formed. Since the parameters of all the environmental objects are known, a path planning algorithm that uses empiric data under the conditions of fully formalized environment can be used. For this purpose the work [14] proposes an algorithm with a flow chart presented in Fig. 2.49.

The following main types of convolution filters are recommended for usage in convolution layers of CNNs.

Digital filters are used to increase the quality of the bitmap by eliminating the false data or improving the data characteristics. These convolution filters are applied to moving or overlapping kernel (window or neighborhood). The convolution filters work by calculation of the pixel value using the weighting neighbors.

There are many types of convolution filters based on different values of kernels. Some of them are intended for smoothing the image, others can be used to highlight and emphasize the image boundaries (edges). It is also possible to join the filters to get certain results. For example, we can apply a filter that can eliminate a stain or smooth the image. Then an edge filter can be applied.

To get a better representation it is recommended to apply a histogram stretching to adjust the contrast and brightness of the image helping to find the spatial objects.

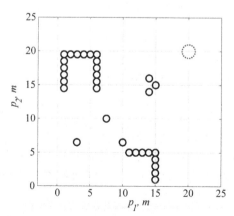

FIGURE 2.48 Virtual scene for generation of training samples.

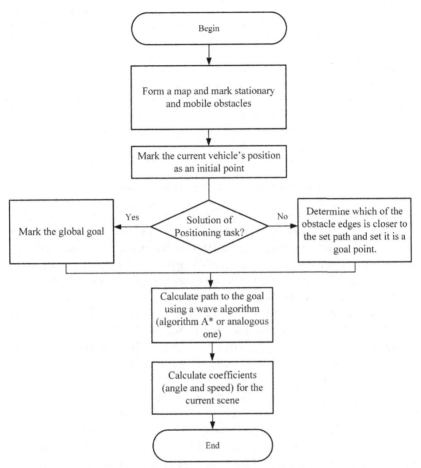

FIGURE 2.49 Algorithm of forming the local paths for generation of training sample.

In order to demonstrate the capabilities of the proposed approach, in Ref. [14], the generated images were split into 25 classes according to the obstacle's distance to the center and the distance from vehicle to the obstacle.

As described in Section 2.5.2 training was split into two stages. At the first stage, the system is trained to classify the input images. In this mode, the network structure presented in Fig. 2.47 is changed. The two fully connected layers, with seven outputs equal to the number of quadratic form equation coefficients, are substituted with one fully connected layer with 25 outputs corresponding to 25 classes of images.

The network in Ref. [14] was trained to work with classification task in five iterations. For each of them 2000 training samples were processed. Since the first training stage is intermediate and its purpose is to form weights capable of searching for invariant features, it is imposed with weaker requirements on classification accuracy. In addition, there is no sense in achieving high accuracy

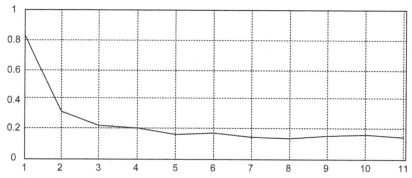

FIGURE 2.50 Graph of classification error changing for the first training stage.

because the last fully connected layer will be removed from the structure at the second stage. A gradient descent algorithm is selected as a training algorithm. The initial global training coefficient is set to 0.005. After each iteration, the coefficient is reduced by a factor of 0.4. The graph of changing the classification error is presented in Fig. 2.50. As a result the classification error was equal to 15%.

At the second training stage the network structure returns to the original state (two fully connected layers) and is trained using the stochastic Levenberg–Marquardt method described in Section 2.5.3.

The plot of root-mean-square error that occurs in the training process using Levenberg–Marquardt method is presented in Fig. 2.51. This plot gives a general view of the NN training.

To obtain more accurate results, the local neural networking planner was applied to a testing sample. The planner functioning quality was estimated using two characteristics: path accuracy and classification quality. The latter was determined as a percentage of false paths crossing the obstacle. The path accuracy was determined as RMSD of the trajectory formed by the NN in comparison to trajectory formed based on the exact information about the obstacle location for the correctly classified paths [14].

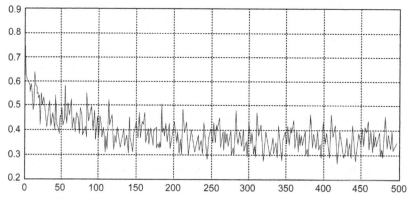

FIGURE 2.51 RMSD plot.

2.5.5 Simplified Version of a Neural Networking Motion Planner Based on a Convolutional Neural Network

Accounting for the parameters of the used locator (see Section 1.5), the structure of the NNPP can be essentially simplified by usage of CNN oriented on processing the two input data vectors holding distance to obstacles and goal position.

It is proposed to use the samples obtained in modeling an FN as training samples. The necessary parameters of the current path are generated at the output of the planner according to formulas (1.9) and (1.10) similar to the previous algorithms. The structure of such planner is presented in Fig. 2.52 and the architecture of the simplified convolutional network in Fig. 2.53.

Since the input data is unidimensional (unlike the classical CNN, where it is represented by a 2D array) any filters with the necessary cutoff frequencies can be used for feature maps. FIR filters are used more frequently than IIR filters due to possibility of building FIR filters with a linear phase response. In addition, FIR filters have a number of merits making them preferable in comparison to IIR filters:

- FIR filters are absolutely stable.
- FIR filters do not require a feedback in their implementation.
- Phase response of an FIR filter can be linear.

There are many filters that can be categorized as follows:

1. Interpolating filters having rather narrow impulse response.
2. Bell-shaped response filters. These filters are good in filtering high-frequency noises.
3. Window sinc-filters are usually ideal for low frequencies but can be implemented due to its infinite length. If the frequency response of a sinc-filter is multiplied by a window function it gives an implementable filter with good spectral qualities.

The main parameter of a filter is its impulse response influencing the calculation of new values during filtration. Variety of features based on usage of different filters can increase the number of feature maps and can have a positive effect on recognition of a certain situation and on generation of the recommended parameters of vehicle's motion. However, if the number is too high, it can have a negative effect on the training time and processing speed.

In order to select the most effective filters one should know the selection criteria and what characteristics are to be analyzed. There are no clear recommendations in the CNN theory and the selection process is empiric and intuitive. However, in practice, according to Refs. [16,45], the following filters are most widely used: leapfrog filter, triangular filter, Hann filter, Hermite filter, Bell filter, bicubic filter, Catmull-Rom filter, cubic spline filter, Mitchell filter, Gauss filter, Lanczos filter, and Blackman filter.

For selection of a particular filter one can use either classical evaluation criteria, such as mean difference, normal correlation, correlation quality, root-mean-square error, signal/noise ratio, or the new integral criteria based

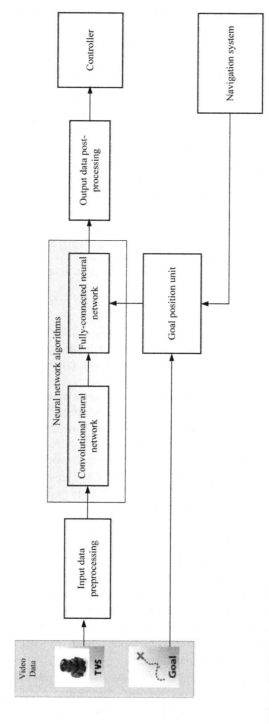

FIGURE 2.52 Operating diagram of a simplified neural networking path planner based on CNN.

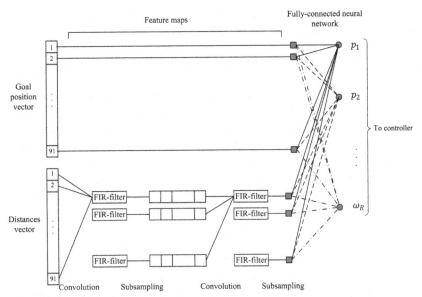

FIGURE 2.53 Architecture of simplified convolutional network for path planning.

on weighted estimate of the filtration quality and specific situation, which can be a topic for a separate research.

This section presents the task of application of CNN for planning the vehicle's motion, describes the general approach and the CNN structure, and filters proposed to be used as feature maps. The conclusions are [14] made concerning their successful application for situation assessment, obstacle detection, and finding the motion direction.

However, generally CNN is a non-formalized structure that is designed intuitively to a large extent. It mostly suits for recognition of situations in real time. The possibility to train CNN in the process of vehicle's motion allows combining this approach with the bionic method of neural networking search proposed before. From our point of view a new symbiosis of these methods will give significant reduction of the probability of repetitions and looping of the vehicle's paths.

2.6 SUMMARY

This chapter presents the results of development and research of methods and structures of intelligent NNPP as a part of intellectual systems of position-path control for robotic vehicles functioning under environmental uncertainty conditions. The considered methods belong to the class of search methods without memory and are based on a bionic approach to intellectual planning systems design.

Based on the research done, a DVH method is proposed together with its neural networking modification (DVH-NN). A bionic method of NNPP is proposed. In real time it ensures dynamic search for the vehicle's motion direction according to the data coming from the lookahead multibeam locator. It also forms the recommended motion parameters necessary for the position-trajectory controller for correct functioning at the next control process step.

The main advantages of the proposed DVH method are its clarity and simplicity, low computational costs, high processing speed, and possibility of dynamic vehicle speed control in passing the dangerous zones. It uses the initial data "as is," i.e., in the form of distance vector (histogram) generated and given out by the locator to the planner. An additional merit of the DVH-NN method, together with bionic method of NNPP, is the possibility of its implementation by the modern FPGAs as reflected in the structural implementations of the neural networking planner presented in this chapter. Hardware implementation of the neural networking planner should lead to an increase of the planning subsystem's efficiency becoming several orders greater than its software implementation. Moreover, bionic method allows online building of the motion path models inside the locator's active zone which allows flexible cohering of the intellectual planning subsystem speed with other functional subsystems of the robot.

The results of the research show that the functioning quality of the proposed algorithms directly depends on the configurational scene complexity, range and type of the installed locator, and the robot's motion speed. Just like for all other "memoryless" search methods, each of the proposed methods can get into a loop. The standard "right hand rule" does not solve this situation at all times. Often, application of this rule results in a nonoptimal path. In addition, the proposed methods have a low value of safety factor. The quality of the decision made depends on the amount of obstacle's angular widening and accuracy of determining the vehicle's and goal's coordinates in the process of their mutual motion.

From the author's point of view, the safety factor can be increased by means of integration of the proposed approach with the potential field methods. Joining the developed neural networking planners with the CNN will reduce the probability of vehicle's path looping and will increase the decisions quality for the conditions of partial environmental uncertainty. The CNNs should be taught in advance the typical situations emerging in the assumed vehicle's functioning environment. CNN can be fine-tuned during the vehicle's motion. For this reason, the final part of the chapter pays a lot of attention to the problem of applying CNN to planning the vehicle's motion and description of the general approach to CNN's structural organization. It is suggested to use digital filters in forming the feature maps for situation assessment, obstacles detection, and finding the recommended motion direction. The CNN teaching methods are also considered.

REFERENCES

[1] P.K. Anohin, General Theory of the Body Functional Systems, Progress of Biological and Medical Cybernetics, Medicina, Moscow, 1974, pp. 52–110.
[2] Y.V. Chernuhin, Micro-Processor and Neural-Computer Control of Adaptive Mobile Robots: Tutorial, TSURE, Taganrog, 1993.
[3] Y.V. Chernuhin, Y.S. Dolenko, P.A. Butov, Bionic approaches to the proceeding of sensor information in neural networks control systems of intelligent mobile robots, Izvestiya SFedU, Engineering Sciences 5 (2012).

[4] A.S. Devyatisylny, I.B. Kryzko, Stabilization of unmanned object motion along the programmed trajectory, proceedings of the Russian Academy of Sciences, Control Theory and Systems 4 (1995) 228−233.

[5] A.S. Yushchenko, Robot control using fuzzy logics: state and problems, News of the Artificial Intellect 1 (2006) 119−130.

[6] A material from Wikipedia—free encyclopedia. https://en.wikipedia.org/wiki/Convolutional_neural_network.

[7] Convolutional Neural Networks (LeNet). Deep Learning 0.1 Documentation. Deep Learning 0.1. LISA Lab.

[8] J. Borenstein, Y. Koren, The vector field histogram − fast obstacle avoidance for mobile robots, IEEE Journal of Robotics and Automation 7 (3) (1991) 278−288.

[9] V.F. Guzik, V.V. Kurejchik, M.Y. Polenov, Intelligent Systems, Issue 6, SFEDU, Rostov-on-Don, 2013.

[10] V.V. Kruglov, V.V. Borisov, Artificial Neural Networks, http://www.socioego.ru/teoriya/istoch/neyron/sod.html.

[11] Y.V. Chernuhin, Neuroprocessor Networks, TSURE, Taganrog, 1999.

[12] D. Du Bois, A. Prad, Theory of Possibilities, Translation from French by V.B. Tarasova, Radio and Communications, Moscow, 1990.

[13] XII All-Russian Conference for Control Problems: Analytical Review, 2014. http://vspu2014.ipu.ru/taxonomy/term/101.

[14] V.K. Pshikhopov, M.Y. Medvedev, R.V. Fedorenko, Aeronautical Complexes' Control: Design Theory and Technology, Fizmatlit, Moscow, 2010.

[15] V.K. Pshikhopov, M.Y. Sirotenko, Structural-algorithmic realization of control system of the autonomous mobile robot with neural network planner, in: Proceedings of TSURE, Special Issue "Intelligent CADs", Taganrog, vol. 3, 2004, pp. 185−191.

[16] M.Y. Sirotenko, V.K. Pshihopov, Building-up principle for mobile robots neural network motion planners for a priory non-formalized environments, Izvestiya SFedU, Engineering Sciences 1 (2008) 196−198. Taganrog: SFEDU.

[17] Y. LeCun, Y. Bengio, Convolutional Networks for Images, Speech, and Time Series, Handbook of Brain Theory and Neural Networks, MIT Press, Cambridge, Massachusetts, 1995, pp. 255−258.

[18] A. Marin-Hernandez, M. Devy, V. Ayala-Ramirez, Visual planning for autonomous mobile robot navigation, in: MICAI 2005: Advances in Artificial Intelligence, vol. 3789, Springer, Berlin−Heidelberg, 2005.

[19] D. Burschka, J. Geiman, G. Hager, Optimal landmark configuration for vision-based control of mobile robots, in: Proc. of IEEE ICRA, Taipei, Taiwan, September 14−19, 2003, pp. 3917−3922.

[20] J.B. Hayet, F. Lerasle, M. Devy Visual, Landmarks detection and recognition for mobile robot navigation, in: Proc. of IEEE Conf. on Computer Vision and Pattern Recognition (CVPR'2003), vol. 2, 2003, pp. 313−318. Madison, Wisconsin, USA.

[21] F. Jurie, M. Dhome, Hyperplane approximation for template matching, IEEE Transactions on Pattern Analysis and Machine Intelligence 24 (7) (2002) 996−1000.

[22] R. Möller, Perception through anticipation, an approach to behaviour- based perception, in: Proc. of New Trends in Cognitive Science, Vienna, 1997, pp. 184−190.

[23] K. Klein, V. Sequeira, View planning for the 3D modelling of real world scenes, in: IEEE/RSJ IROS, vol. 2, 2000, pp. 943−948.

[24] K.A. Tarabanis, R.Y. Tsai, A. Kaul, Computing occlusion-free view- points, IEEE Transactions on Pattern Analysis and Machine Intelligence 18 (3) (1996) 279−292.

[25] N.J. Nilsson, Shakey the Robot, Technical Note 323, SRI International, 1984.

[26] I. Horswill, Visual collision avoidance by segmentation, in: Proc. of the IEEE, RSJ Intern. Conf. on Intelligent Robots and Systems, 1994, pp. 902−909.

[27] M.A. Turk, M. Marra, Color road segmentation and video obstacle detection, in: SPIE Proc. of Mobile Robots, vol. 727, 1986, pp. 136−142. Cambridge, MA.

[28] M.I.A. Lourakis, S.C. Orphanoudakis, Visual Detection of Obstacles Assuming a Locally Planar Ground, Technical Report FORTH-ICS, TR-207, 1997.

[29] L.M. Lorigo, R.A. Brooks, W.E.L. Grimson, Visually-guided obstacle avoidance in unstructured environments, in: Proc. of IEEE Conf. on Intelligent Robots and Systems, 1997, pp. 373−379.

[30] H. Zhang, J.P. Ostrowski, Visual motion planning for mobile robots, IEEE Transactions on Robotics and Automation 18 (2) (2002) 199−208.

[31] D. Coombs, K. Roberts, 'Bee-bot': using peripheral optical flow to avoid obstacles, in: Intelligent Robots and Computer Vision, Boston, MA, vol. 1825, 1992, pp. 714−721.

[32] D. Coombs, M. Herman, T. Hong, M. Nashman, Real-time obstacle avoidance using central flow divergence and peripheral flow, IEEE Transactions on Robotics and Automation (1995) 276−283.

[33] J. Santos-Victor, G. Sandini, Uncalibrated Obstacle Detection Using Normal Flow, 1995.

[34] J. Santos-Victor, G. Sandini, F. Curotto, S. Garibaldi, Divergent stereo in autonomous navigation: from bees to robots, International Journal of Computer Vision (1995) 159−177.

[35] Y. LeCun, F.J. Huang, L. Bottou, Learning methods for generic object recognition with invariance to pose and lighting, in: Proc. of CVPR'04, IEEE Press, 2004.

[36] S.C. Sukittanon, A.C. Surendran, J.C. Platt, C.J. Burges, Convolutional Networks for Speech Detection, 1995.

[37] Y. LeCun, J.S. Denker, S. Solla, et al., Optimal brain damage, in: Advances in Neural Information Processing Systems 2 (NIPS'89), Morgan Kaufman, Denver, CO, 1990.

[38] F.J. Huang, Y. LeCun, Large-scale learning with SVM and convolutional nets, in: Proc. of Computer Vision and Pattern Recognition Conference (CVPR'06), IEEE Press, 2006.

[39] Y. LeCun, L. Bottou, Y. Bengio, P. Haffner, Gradient based learning applied to document recognition, Proceedings of the IEEE 86 (11) (1998) 2278−2324.

[40] C. Neubauer, Evaluation of convolutional neural networks for visual recognition, IEEE Transactions on Neural Networks 9 (4) (1998).

[41] Y. LeCun, Generalization and network design strategies, connectionism in perspective, in: Proc. of the Intern. Conf. Connectionism in Perspective. Univ. of Zurich, Elsevier, Amsterdam, 1988.

[42] H. Saimon, Neural Networks, Complete Course, Williams, Moscow, 2006.

[43] F. Wasserman, Neuro-Computer Equipments: Theory and Practice, Mir Publishers, Moscow, 1992.

[44] Y. LeCun, L. Bottou, G. Orr, K. Muller, Efficient BackProp, Neural Networks: Tricks of the Trade, Springer, 1998.

[45] V.K. Pshikhopov, M.Y. Sirotenko, B.V. Gurenko, Structural Organization of Submersibles Automated Control Systems for a Priori Non-Formalized Environments, Information and Measuring and Control Systems, in: Intelligent and Adaptive Robots, vol. 4, 2006, pp. 73−79. Moscow.

[46] J.G. Nicholls, A. Robert Martin, P.A. Fuchs, D.A. Brown, M.E. Diamond, D. Weisblat, From Neuron to Brain, 2012.

[47] K. Fukushima, Neocognitron: a self−organizing neural network model for a mechanism of pattern recognition unaffected by shift in position, Biological Cybernetics 36 (4) (1980) 193−202.

Chapter Three

Vehicles Fuzzy Control Under the Conditions of Uncertainty

E. Kosenko, D. Beloglazov, V. Finaev

Southern Federal University, Taganrog, Russia

3.1 TYPES OF UNCERTAINTIES

The vehicle control algorithms considered in Chapter 1 [1] can be effectively used in certain conditions with adequate description of the environment. The real functioning conditions are far from satisfying these requirements. Hence, it can be undoubtedly declared that the vehicle functions under the conditions of partial initial data and uncertainty of environmental parameters.

Any mobile vehicle control is a process of generation of options, decision making, and implementation of the decisions based on the analysis of data about the vehicle and the environment. If the environment is certain, automatic control theory [2,3] is used in technical systems, otherwise incompleteness of data makes the choice of decision theory more preferable [4,5].

The types of uncertainties can be categorized as shown in Fig. 3.1.

The first level of uncertainties tree is formed by the terms Obscurity, Unreliability and Ambiguity giving a qualitative measure of the amount of information missing in the description of the control problem elements.

The second level describes the sources of ambiguity represented by the environment (physical) uncertainty and by the person taking the decisions and professional language (linguistic uncertainty). Based on the classification of uncertainties (Fig. 3.1), the research of control problems solution methods, under the conditions of data incompleteness, falls into the category of uncertainties connected to inaccuracies and fuzziness caused by the environmental action, presence of aftereffect, unsteadiness, and influence of decision-making persons.

The application sphere of control problems under the conditions of uncertainty includes the *intelligent systems of position-path control* [6,7] functioning under the conditions of constantly changing situation, presence of multiple disturbances, and arbitrary actions, which makes it necessary to take adequate solutions under the conditions of data incompleteness.

97

Path Planning for Vehicles Operating in Uncertain 2D Environments. http://dx.doi.org/10.1016/B978-0-12-812305-8.00003-X

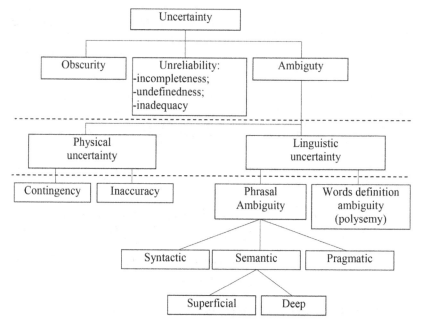

FIGURE 3.1 Classification of uncertainty types.

Taking optimal control solutions is one of the most complicated control problems under the conditions of uncertainty and partial knowledge about the control consequences. It requires application of system approach [8] and original mathematical models [9] for the system research of control problems.

Formalization of the decision-making tasks under the conditions of uncertainty calls for necessity to use the methods of fuzzy sets theory [10–13], possibility theory [14], and situational models of decision making [15–17].

Systems functioning in fuzzy environment with parameters defined verbally (presence of fuzzy goals and limitations) are called fuzzy-goal systems [18]. Research methods based on mathematical decision theory generally allow the solution of analysis and optimization tasks of fuzzy-goal systems but require further development (extension), because of the specific character of the fuzzy-goal control requiring application of methods, directed on activation of intuition and experience of specialists, and building control systems models using system parameters in the form of fuzzy intervals.

The main problem of autonomous vehicle's motion is the presence of many uncertainties concerning the state of the environment.

The fuzzy-logic theory allows reduction of the uncertainty and data incompleteness. This approach is based on human capability to process information in perception. The rules constituting the fuzzy logic ensure formal methodology for the linguistic rules ensuing from the reasoning and decision making based on indistinct and incomplete information.

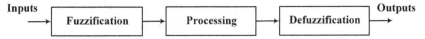

FIGURE 3.2 Control sequence using fuzzy logic.

The logical fuzzy control outputs are obtained as a sequence of the following steps: fuzzification, processing, and defuzzification, as shown in Fig. 3.2.

The fuzzification unit defines, for example, a fuzzy set A in a universal set X by its membership function $\mu_A(x)$ defining the degree of membership of x in X for each x. In fuzzy control, the membership functions quantify the grade of membership of values of physical variables to linguistic terms.

Then in the processing unit using the fuzzified values of input variable a decision-making task is solved using the base of fuzzy rules. This rules base is used for description of relation between fuzzy inputs and outputs. For example, a simple fuzzy control rule connecting input v and output u can be expressed as a conditional operator in the following form:

$$\text{IF } v \text{ is } W \text{ then } u \text{ is } Y \tag{3.1}$$

where W and Y are the fuzzy values defined in the areas v and u.

Inference for each fuzzy rule is defined according to the membership value of the input variables. The mechanism of logical inference generates a set of control actions according to the fuzzified input values.

Since the control actions are presented in a fuzzy form, the defuzzification method should transform these fuzzy control actions into specific real values for the fuzzy controller.

Usually, in order to determine the behavioral strategy for the navigation problems, the main behavior is split into a number of simpler ones (independent behaviors). For the fuzzy-logical behavior, each behavior is put together using a set of fuzzy rules intended to reach a well-defined set of goals. For example, a rule can be defined as follows:

*If the goal is close and to the left, **then** turn left and move with a low velocity.*

3.2 APPLICATIONS OF FUZZY LOGIC IN VEHICLES CONTROL

Autonomous robot, as a vehicle, should be created as a class of intelligent technical systems having a purposeful behavior capable of spatial motion and performing the necessary actions. The modern robots are equipped with technical vision systems and sensors in order to get the information about the current position in space. The robot's knowledge base allows it to make decisions about the actions necessary to perform the set task.

During supervisory control, a human operator gives robot a task using a "problem-oriented language" and can observe the robot's actions. Operator receives the robot's messages giving him the feedback, so that he can correct the commands, and informing the operator about the current situation or reaching the goal. The supervisory control task assumes a certain level of "rapport," when a human and a robot use the same space–time assessments in scene analysis, reasoning schemes, and logical deductions that are understandable for the human.

Development of the supervising control for the purposeful action for robotic vehicles is performed using the fuzzy sets and fuzzy logic [10–14]. Theoretical foundations of describing the external world using the fuzzy and natural space–time relations were laid in the works of D.A. Pospelov and his coauthors [16,19].

This approach should develop further in order to get the most adequate estimations of the situations, take control decisions, and ensure natural behavior of the robot in a priori undefined conditions.

3.2.1 Formalization of Environmental Description

The main tasks of vehicle's purposeful fuzzy control theory are the description of the external world and the current situation on the verbal level using the linguistic variables (LVs) and fuzzy variables (FVs), planning operations, and organization of operator's conversational interface.

Description of the vehicle's environment includes description of the objects in the space of vehicle's functioning, spatial relations between the environmental objects and the vehicle.

The spatial relations between the vehicle and the environmental objects can be described as both extensional and intentional [20].

Extensional relations are used for formalization of position and orientation of objects. For example, in order to describe the relations between two objects, one can use binary orientational relations including verbal elements: f_1 is the object a_1 ahead of object a_2, analogously; f_2, to the left and ahead; f_3, to the left, etc.; distances: d_1, right up; d_2, close; d_3, not close–not far, d_4, far; d_5, very far. The fuzzy relation graph is defined using the membership functions that are usually determined experimentally and account for the human perception of spatial relations.

Fig. 3.3 [20] shows the example of setting the membership function for the LV "distance" with d_i is built using experimental data.

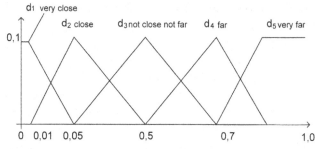

FIGURE 3.3 Membership functions for the linguistic variable "distance" built using experimental data.

Intentional binary relations define the object formally, e.g., in the form of a relation with the following elements: R_1, touch; R_2, be inside; R_3, be outside; R_4, be at the center; R_5, be in line; R_6, be at one surface; R_7, have a nonzero projection; and R_8, stand at the surface.

The work [19] considers formalization of two unary relations—R_{00} be horizontal and R_{01} be vertical, and 28 elementary spatial binary relations. These relations are obtained from the basic relations using the formal logical rules of conjunction and disjunction.

The union of set objects in the motion space of the mobile object, and the multitude of relations between them together with the transformation rules, constitute the formal language for situation description. As a result, the situation description language using the logic of space—time relations provides a semiotic formalization of the mobile object and the environmental objects. For example, the spatial relation "a_1 stands on surface S far to the right" is formally defined as follows:

$$(a_1 R_8 S) \& (a_0 d_5 f_7 a_1)$$

where a_0 is the observer with respect to whom the relations of distance and orientation are formulated for the object a_1.

The state of the environment of the mobile object as a current situation is described by the system of binary frames (<object m>, <relation>, <object $m + 1$>), $m = 1, 2, ..., M$, where one of the objects can be a mobile object or an external observer.

If fuzzy relations are set between all the spatial objects that can be observed by the vehicle during its motion, we get a fuzzy semantic network or a "fuzzy map." Application of a fuzzy map allows us to solve the task of navigating the mobile object using the observed references—objects with a known position. Determining its position using the references, the vehicle can use the fuzzy relations to calculate its own position and the direction to the goal, knowing its coordinates. Introducing the third fuzzy coordinate for elevation (h_1, at level; h_2, higher; h_3, much higher; h_4, below; and h_5, much below) and using 3D relations between the objects, the vehicle can plan its motion in space accounting for the landform [20].

The vehicle analyzes the state space using obstacles identification algorithms that constitute a system of fuzzy-logic inference using a set of situation classification rules and fuzzy-object features creating a knowledge base representing the obstacles.

Mamdani algorithm [21,22] can be used for a fuzzy classifier. The object's fuzzy parameters (length, width, height, dimension, etc.) are defined by the membership functions of FVs forming a term-set of the corresponding LVs. The membership functions are defined on some basic sets that are, in turn, defined by the technical characteristics of the vehicle and its sensory system. These are defined before the motion is started. Using the fuzzy features [20,23], the obstacle's nature and its parameters are determined with the help

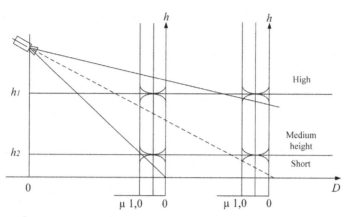

FIGURE 3.4 Membership function in relation to the sensor distance.

of environment analysis algorithms, the preset membership functions, fuzzy production rules, and the parameters of the real situation.

When controlling objects such as a mobile robot, there is a peculiarity in application of its sensory system related to the fact that the image perceived by the technical vision system changes its scale during the motion.

The work [20] proposes to introduce a two-dimensional membership function. For example, for the terms of the LV "*Object height*" the membership functions are introduced accounting the mounting angle of the vision system. The relation of the membership function from the distance to the sensor is shown in Fig. 3.4. The orientation membership functions ("to the left," "to the right") also change according to the distance.

The state of the space objects can change in time as they move. So the situation description also changes in time reflecting the vehicle's motion. This requires accounting for not only for spatial but also time relations in the motion space, e.g., *earlier or later, simultaneously, follow*.

Therefore, the state in the motion space of the vehicle S is defined by a frame with its slots being the names of the space objects, and spatial and time relations between the objects.

3.2.2 Formalization of the Vehicle's Behavior

The vehicle's behavior is its motion in space intended to reach the goal point. In order to formalize the behavior and decision making, various situations [15−17,24−27] can be used. A common feature of the models is the presence of situational analysis. The experts set the logical rules of fuzzy inference for the decision to be taken. The expert knowledge is formalized in the description of environmental objects and the vehicle itself. Let us consider the most common types of models.

At the conceptual level defining the motion model means finding an operator W that in set-theoretical approach serves as a relation between the input parameters, environmental states, and output parameters of the vehicle.

If X, input parameters vector, Y, output parameters vector, B, vehicle design parameters, in set-theoretical approach, the operator W is a relation between the input parameters vector X, state vector B and output parameters vector Y. The vehicle motion model is defined by the following form:
- transition functions

$$p = \langle X, B, P \rangle, \; X \times B \xrightarrow{P} B \qquad (3.2)$$

where P is the correspondence graph of p;
- outputs functions

$$f = \langle X \times B, Y, F \rangle, \; X \times B \xrightarrow{F} Y \qquad (3.3)$$

where F is the correspondence graph of f.

Any model of vehicle's behavior and decision making requires determining the type and form of setting the correspondence graphs P and F, application of fuzzy inference, and algorithmization.

CLASSIFICATION MODEL

Input variables of the vehicle's behavior are defined in the form of LVs $\alpha_i, i = \overline{1,n}$ and FVs $\alpha_i^k, k = \overline{1,m}$ that constitute term-sets LV α_i.

The experts define the set of controlling solutions $H = \{h_1, h_2, ..., h_m\}$ and formulate the decision-making rules represented by the corresponding table for the collections of FV and elements of the solution set.

The decision-making model is defined as a following set

$$(X, \Psi, H)$$

where X is a set of features, i.e., state factors of the vehicle; Ψ decomposition of X into fuzzy reference classes L_j, $\left(j = \overline{1, |H|} \right)$. The set $X_i \subset X, i = \overline{1,n}$ is a basic set for defining the LV α_i and its FV from the term-set $T(\alpha_i)$.

As a result, a table "situation-action" is formed that formally defines all the possible situations at the verbal level and assigns the corresponding control solutions.

For each combination of FVs $\alpha_i^k, k = \overline{1,m}$ and LVs $\alpha_i, i = \overline{1,n}$ experts formulate a decision-making rule and a fuzzy modus ponens rule [28]. The total number of rules is equal to $|T(\alpha_1)| \times |T(\alpha_2)| \times ... \times |T(\alpha_n)|$. The reference classes corresponding to the same solution are selected out of all the multitude of rules.

For each reference class a membership function is defined for jth solution belonging to jth reference class.

$$\mu_{L_j}(x_1 x_2 ... x_n) = \bigvee_{\left(\alpha_1^j, \alpha_2^j, ..., \alpha_n^j \right) \in L_j} \mu_{\alpha_1^j}(x_1) \& \mu_{\alpha_2^j}(x_2) \& ... \& \mu_{\alpha_n^j}(x_n),$$

$$x_i \in XI, \quad i = \overline{1,n}, \; j = \overline{1, |H|} \qquad (3.4)$$

where n_j is the number of sets $\alpha_1^i, \ldots, \alpha_n^i$ belonging to jth decomposition class.

A control solution specifying the vehicle's future behavior is selected in the following way. The physical values of the state parameters for the vehicle and the motion space objects $\left(x_1^0, x_2^0, \ldots, x_n^0\right) \in X$ are determined for the decision-making moment. These values are substituted into the FV membership functions. The values of the membership degrees for the reference classes $\mu_{L_j}\left(x_1^0, x_2^0, \ldots, x_n^0\right)$, $j = \overline{1, |H|}$ are calculated, and a maximum value μ_{L_s} is found among all the values of μ_{L_j}

$$\mu_{L_s} = \max_j \mu_{L_j}\left(x_1^0, x_2^0, \ldots, x_n^0\right) \tag{3.5}$$

A control solution h_s with an index s is considered to be the most suitable for the considered situation and is accepted with a membership degree μ_{L_s}.

MODEL FOR CALCULATING THE TRUTH DEGREE OF THE FUZZY PRODUCTION RULES

The model is defined by the following triad:

$$X \times H \xrightarrow{T} H \tag{3.6}$$

where T is the fuzzy relation on a set $X \times H$. The set H is a set of FVs from the term-set of LV "control solution." The experts perform the selection of the T elements in a form of fuzzy production rules $\{\pi_j\}, j = \overline{1, l}$ that are formalized by conditional and unconditional operators. For each statement π_j a membership function can be defined

$$\mu_{\pi_j}(x_1, x_2, \ldots, x_n, h_i)$$

For the relation T a membership function is defined by a generalized operation σ so that

$$\mu_T(x_1, x_2, \ldots, x_n, h_i) = \sigma_{\mu_{\pi_j}}(x_1, x_2, \ldots, x_n, h_i) \tag{3.7}$$

A control decision is taken in the following way. At the decision-making moment t_0, the state coordinates of the vehicle and the motion space objects are determined as $x^0 = \left(x_1^0, x_2^0, \ldots, x_n^0\right) \in X$. For the point x^0 the values of the membership functions $\mu_{T(\pi_j)}\left(x^0, h_i\right)$ are found; the functions of fuzzy-logical selection of the control solution h_i that depends on the values of the membership degrees of the solutions μ_{h_i}. A selected solution is such a value of the basic set of LV "control decision" that gives a maximum value to the membership function $\mu_{T(\pi_j)}\left(x^0, h_i\right)$

$$\mu_{T(\pi_s)}\left(w^0, h_s\right) = \max_j \mu_{T(\pi_j)}\left(w^0, h_i\right) \tag{3.8}$$

Note that the completeness of the set T defined by the experts has a major influence on the model's performance and validity.

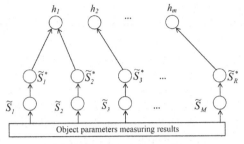

FIGURE 3.5 Situational model of control decisions making.

SITUATIONAL MODEL
In the situational model, the fuzzy-logical production is the process of selecting a control solution based on the analysis of real fuzzy situation in the vehicle's motion space and their correlation with the reference fuzzy situations, set by the experts in such a way that each fuzzy situation is matched with a certain control solution.

In Fig. 3.5 the vehicle's model is presented in the form of a graph.

$S^* = \{\tilde{S}_1^*, \tilde{S}_2^*, ..., \tilde{S}_R^*\}$—fuzzy reference situations;

$S^* = \{\tilde{S}_1, \tilde{S}_2, ..., \tilde{S}_M\}$—fuzzy real situation;

$H = \{h_1, h_2, ..., h_m\}$—control solutions set.
These fuzzy situations are defined as type-2 fuzzy sets [29]:

$$\tilde{S} = \left\{ \left\langle \frac{\mu_s(\alpha_i)}{\alpha_i} \right\rangle \right\}, \alpha_i \in A$$

where $\alpha_i, (i = \overline{1, n})$ ith LV.

In order to produce a control solution, the real fuzzy situation $\tilde{S}_i$ is compared to the fuzzy reference situations $\tilde{S}_j^*, j = \overline{1, R}$ using the operations of fuzzy logic, i.e., determining the degree of fuzzy equality. If all the values of fuzzy equality degree are lower or equal to 0.5, there is uncertainty that can be avoided by calculating the value of fuzzy degree of commonality between situation $\tilde{S}_i$ and situations $\tilde{S}_j^*, j = \overline{1, R}$, or by application of additional research.

A common point for all the considered models of vehicle behavior and decision making is that for the basic sets $X_1, X_2, ..., X_n$ the experts set the degrees of FVs membership in the basic sets along with the decision-making rules. For the decision-making time t_0 they define the factor coordinates for the motion space as

$$\left(x_1^0, x_2^0, ..., x_m^0\right) \in X_1 \times X_2 \times \cdots \times X_n = X$$

and the vehicle itself. The values of membership degrees for the FVs are also defined. Each of the models has its advantages and drawbacks. A decision concerning their application should be made after the final decision-making task statement for the conditions of uncertainty.

Application of the situational model allows marking out the vehicle's stereotypes of behavior in the existing situation. That is, we can apply the notions of the conditional-reflex behavior of the type "stimulus-response." However, the behavioral stereotypes also have the form of production rules: "if situation is S_i, the tactic is T_i."

In many works under the conditions of uncertainty, the tactic is understood as a group of behavioral rules directed toward reaching the goal set for the vehicle. Generally, the decision-making rules are the inference rules that match a typical situation (generally uncertain) with a preset vehicle's motion in space.

The typical (reference situations) are defined by the experts in the fuzzy knowledge base of the vehicle.

It should be mentioned that non-fuzzy values are a special case of fuzzy ones, so the algorithms describing the vehicle's behavior can include both: LVs and non-fuzzy parameters—specific numeric values such as D, distance to the obstacle, W, obstacle's width, H, obstacle's height, etc.

Behavior of the mobile object is generally determined by the frame: <*current situation S_i*> <*plant* a_0> <*operation name*> <*accompanying objects j*> <*operation feasibility conditions*>. The frame defines a plant (vehicle) with its parameters present in the data base (dimensions, mass, power of the movers, speed, maneuverability, etc.). The frame includes the feasibility conditions for the motion operation accounting for the current situation and parameters of the environment such as landform, soil quality, wheel traction, obstacles character.

Another important task is the task of robot's behavior self-analysis under the conditions of uncertainty. This can also be performed using fuzzy-logic methods based on the sensory system's data and the appropriate informational support.

As mentioned earlier in the application of situational model comparison of the observed situation to the reference ones located in the knowledge base is performed using certain fuzzy criteria of situation similarity. The situation estimation received by the vehicle is a formalized and averaged estimation of an analogous situation by a human.

An identification example for a situation of vehicle's autonomous motion for a medical microrobot can be found in the work [30]. The system of the robot's technical vision for a microrobot not only determines the motion tactics in a deformed blood vessel but also forms a preliminary conclusion about its condition for the surgeon-operator.

The robot's knowledge base can include a multitude of behavioral tactics that are selected by the vehicle depending on both the situation and the operator commands, that includes two elements <*operation name*> <*accompanying objects j*>, e.g. <*pass*> <*threshold P*>. In this case formalization of the current situation and checking the feasibility conditions for the operator's command should be performed by the vehicle's intelligent control systems itself.

If the operator's command is incorrect, the vehicle's intelligent system sends a request to the operator. So organization of the robot-operator dialogue becomes a necessity for the vehicle's intelligent control system.

3.3 VEHICLE'S PATH PLANNING

The operator set's the goal to reach only the final point in its functioning space. Therefore, there is a task to plan the vehicle's autonomous motion.

If the space has obstacles with known parameters, the vehicle can calculate an optimal trajectory using the accepted performance criteria (criteria function) facilitated by the presence of the conditional-reflex stereotypes mentioned in description of the vehicle's situational model. However, if the obstacles' coordinates change randomly, the motion path-planning task becomes essentially complicated. This limits the ability to use automatic control.

When solving the planning task using the artificial intelligence methods for the vehicle motion control, there is a possibility to perform continuous comparison of the actual situation observation results (by measuring) to the conditions set in the planning procedure.

This comparison can be implemented at the level of linguistic description of the actually observed situation and the assumed one (existing just in a form of a statement) set by the experts as a reference in a form of statements or a type-2 fuzzy set. For the unsatisfactory comparison results (contradictions), there must be a plan for their elimination so that a desired situation is reached and the vehicle's path is optimized.

Hence for the vehicle's intelligent system, it is necessary to create a bank of typical operations and a possibility to generate other typical operations based on the comparison of the real and assumed situations as proposed in the work [31]. This type of planning task is especially important if the vehicle's path is searched under the conditions of uncertainty and unstationarity.

In planning complicated operations, a multi-step procedure of disambiguation is introduced. First, the target and the actual situation are compared. If they do not match, contradictions are explored and the actions to be performed for their elimination are determined. The preconditions of permissive actions are then checked. They can also contradict with the actual situation and generate new actions. This continues until preconditions for at least one of the permissive actions are satisfied. Later, this operation is performed again (for now, at the planning level) and a new situation emerges. Then, it is analyzed analogously.

As a result, a chain of cohered elementary operations is generated. If the necessary conditions are satisfied, i.e., if the task has a solution, this chain leads to the goal. The procedure can be presented in a form of a directed graph with a goal located at its root [31].

A drawback of the approach presented in the work [31] is that the operator has to define the conditions and rules for the system working in regular and emergency situations that have to be known in advance. In practice, it is rather hard to accomplish for the vehicles functioning in unpredictable situations.

When the rules base is composed for control decision-making system of a vehicle, it is advisable to perform the teaching of the intelligent control system using a "training with teacher" principle [32] or using the trainable fuzzy (hybrid) neural network where control, as information, is memorized and treated by the neural network.

If an obstacle requires tuning of the network parameters, there must be a mechanism of obstacle-type recognition, where a method of cluster analysis of situation parameters can be used. It allows to single out the most probable of the possible known situations.

New emerging situations require additional training of the intelligent control system imposing limitations on the method for the cases when the possible situations cannot be defined beforehand.

Accumulation and procession of the experience is performed by a special type of neural networks. The system includes a unit called "emotions unit" by the authors. It allows to estimate success or failure of a certain action. The future actions are performed using the "accumulated knowledge" from the previous experiments.

Planning of the robot's behavior can be performed using the gnostic types of operations ensuring the missing information is acquired, allowing to relieve the operator from the necessity to analyze the feasibility conditions.

If the slot corresponding to the feasibility conditions is empty, the intelligent control system should determine if these conditions are satisfied. For example, the vehicle can search for the shortest path on the map, determine the possibility of passing according to the dimensions and so on. Such gnostic operations should be present in the knowledge base of the intelligent control system or set by the operator in a dialogue.

Therefore, the fuzzy-logic and situational control methods allow solution of a rather wide circle of tasks connected to vehicles control under the conditions of uncertainty. However, a human participation is needed for the task statement, motion control and in case of emergency situations.

If the systems becomes highly intelligent the operator's role comes down to controlling the behavior of the vehicle by means of setting the tasks.

3.4 DEVELOPMENT OF THE VEHICLE'S BEHAVIORAL MODEL USING FUZZY-LOGIC APPARATUS

3.4.1 Analysis of the Motion Paths Formation Principles

The main task of a vehicle control system is to ensure motion toward the set goal point avoiding undesired and potentially dangerous interaction with other objects.

The work [33] recommends to split algorithms designed for the vehicle to avoid the obstacles into several classes: hypothesis-test, penalty function, skeletonization method; fuzzy logic; neural networks; genetic algorithms.

The method of hypothesis and test consists of three main steps: hypothesis proposition concerning the prospective way between the initial and final point of the vehicle's path; checking the motion directions set for the collisions possibilities; and finding the detour based on the data obtained in obstacle exploration in case of possible collision. This process is repeated until the goal is reached.

The penalty function method is based on assigning each of the vehicle's path variants a certain weight (penalty degree), depending on the obstacles present on the path.

The skeletonization algorithms reduce the space map to a one-dimensional representation simplifying the planning task. Such a representation with a smaller number of dimensions is called a configurational space skeleton.

The fuzzy algorithm is a group of ordered fuzzy rules determining the behavior of the mobile vehicle and including the notions formalized by the fuzzy sets.

All the fuzzy-logical systems function using the same principles as follows:

- fuzzification of the sensor measurements;
- generation of control using the rules base; and
- defuzzification of the output signal and its future use as a simple digital signal.

Fuzzy logic represents the numeric values for linguistic terms, based on which it helps to form effective solutions under the conditions of uncertainty and incompleteness of data concerning the environment. This makes it an effective tool for implementation of control systems for mobile objects. Fuzzy logic was initially developed as an instrument for working with uncertain data, which gives it advantages in practical application compared to the other methods discussed in this chapter.

Fuzzy control systems function based on the base of rules of a knowledge base [34], containing a set of fuzzy rules *if-then* and based on the key knowledge in a specific area or expert knowledge.

The simplicity of the fuzzy systems and the capability to solve a wide circle of tasks without explicit calculations and measurements made it popular among the scientists and researchers. This became a reason for intensive use of the fuzzy-logic apparatus in implementation of vehicle control systems.

Application of fuzzy logic for implementation of vehicle control systems assumes solution of the following tasks:

- forming of the fuzzy sets for representation of positions and forms of environmental objects;
- planning simple fuzzy behaviors such as obstacle avoidance, goal reaching, and wall following;
- activation of a certain fuzzy behavior (or combination of behaviors) depending on the current state of the environment.

Such an approach allows to eliminate uncertainty and ambiguity of information obtained by the system about the state of the environment. Application of the main approaches of the considered method is presented in works [35–50]. The main idea of implementation of the vehicle's separate behaviors is that the general control task can be presented as a collection of smaller subtasks, which simplifies implementation of the whole system. This allows us to build the control system using the module principle simplifying the general implementation of the separate behaviors and ensuring scalability implemented based on the possibility to add new variants of behavior without increasing the system's complexity.

Let us consider the principles of building the vehicle's control systems based on fuzzy-logic system's algorithms.

3.4.2 Analysis of Methods of Vehicle's Behavior Coordination

During the motion of a vehicle in a non-determined environment, its behavior can be represented in the form of combination of independent behaviors. Selection of a particular behavior can be done according to the current state of the vehicle and the environment. One of the main problems in planning the motion trajectories is the coordination of the independent behaviors according to the situation.

In certain cases there can be contradictions between the independent behaviors, which require creation of a special algorithm (unit) for their coordination. For example, there can be a situation of uncertainty in the choice between such independent behaviors as "wall-following," "obstacle avoidance," and "goal reaching," as shown in Fig. 3.6.

The main task of the coordination unit is the selection of the most suitable independent behavior for generation of control actions given out to the actuators of the vehicle.

Let us consider the approaches to solution of the behavior coordination task.

A switching coordination of behaviors is presented in some earlier works [40,41]. In both publications a Brooks categorization architecture was used together with a priorities structure. An activation recommendation with a highest priority is selected while the recommendations of the other competing behaviors are ignored. In certain situations, this approach leads to ineffective results. For example, if the vehicle can hit the obstacle ahead, the behavior "obstacle avoidance" is selected and the vehicle decides, for example, to turn left to avoid the obstacle, while the goal is located to its right and the behavior "goal reaching" gets a negative result.

Other methods are based on the combination of behaviors ranged according to the predefined weighting coefficients. One of them is the motor schema-based approach presented in [37–39,51]. This method uses potential field for determining the output signals of each schema. All the results (outputs) are joined based on weighted summation (addition of weighting coefficients).

The work [48,49] presents a distributed architecture called DAMN (Distributed Architecture for Mobile Navigation) ensuring vehicle's motion based on a

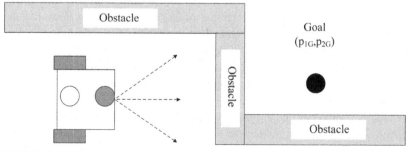

FIGURE 3.6 Conflict of behaviors.

centralized system of "voting" for motion variants belonging to independent behaviors. In this approach each behavior is allowed to "vote" for or against a certain action of a vehicle. The actions that win the voting are performed.

These methods can have low effectiveness. For example, in case of a collision alert, a strategy "avoid obstacle" recommending a left turn is activated. At the same time, the behavior "goal reaching" can recommend motion to the right if the goal is located to the right. As a result, it can lead to a trade-off command resulting in a collision with the obstacle.

For these situations, other approaches that ensure coordination by means of analysis of the vehicle's situation are recommended, i.e., behavior based on a situational context.

In [50,52−56], it is proposed to use a combination of situation-based behaviors with application of fuzzy logic.

In the work [57], a method analogous to situation-dependent combination was developed, where an adaptive hierarchy of several fuzzy behaviors is combined using the concept of applicability degree. In this case, a specific behavior is allowed to affect the overall behavior of a vehicle as required by the current situation and goal.

The problem of forming a context-dependent behavior of a vehicle implies dividing the overall behavior into components, i.e., separate, independent behaviors focusing on execution of a specific subtask. For example, a behavior focuses on reaching the final goal, while another independent behavior focuses on obstacle avoidance. Each behavior is composed of a set of fuzzy-logic rules aimed at achieving a given desired objective. The behavior description consists of a set of fuzzy-logic rules for vehicle's *Velocity* and *Steering* of the following form [43].

$$\text{If } C \text{ then } A. \tag{3.9}$$

where the condition C is composed of fuzzy input variables and fuzzy connectives (And) and the action A is a fuzzy output variable.

So based on Refs. [43,50], a vehicle behavior control system structure is presented in Fig. 3.7.

Instead of processing all the behaviors trying to combine them, the coordination unit analyzes the current situation and takes a decision about the selection of one independent behavior. Such an approach saves time and lowers computational costs.

The main complication in synthesis of fuzzy systems of vehicle's motion planning is the development of control rules bases that are necessary to implement independent behaviors. The base of rules presented in the work [43] allows to plan the vehicle motion path in environments with obstacles of a simple shape.

The rules base does not guaranty the motion safety for the environments with more complicated shapes of obstacles.

So it is proposed to modify this method be extending the control rules base for the separate behaviors and by developing a new way of their coordination.

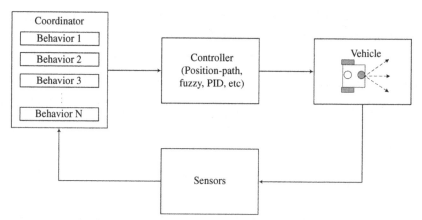

FIGURE 3.7 Architecture of context-dependent coordination of behavior.

3.5 VEHICLE MOTION CONTROL PRINCIPLES

3.5.1 General Motion Control Principle

Solution of the goal-reaching task using the method of situation-dependent co-ordination of behaviors requires decomposition of the overall behavior to a set of simple behaviors represented using *fuzzy if-then* rules [43].

For the vehicle to reach the goal, the following positions are introduced. The robot navigation comprises three behaviors: "Goal-reaching behavior," "Obstacle-avoidance behavior," and "Wall-following behavior." Each behavior is composed of a set of fuzzy-logic rules achieving a precise goal. The output of each behavior represents the *Steering* angle and the *Velocity*. The supervision layer activates independent behaviors depending on the current situation.

The general form of the path-planning algorithm is presented in Fig. 3.8.

The main functions (modules) of the algorithm are executed in a loop until the goal (point in space) is reached.

At each algorithm, iteration vehicle position (p_{1V}, p_{2V}, f_i) is determined, where p_{1V}, p_{2V} are the current coordinates of *x*- and *y*-axes, and f_i is the vehicle orientation angle. Then an array $[X, Y]$ is defined to hold obstacles' coordinates in the radar's visibility range. This array together with the values S_i of the sensor helps us calculate the corresponding distances D_i, $I = 1, ..., 90$ to the detected obstacles. The dimension of I corresponds to the sweep of the radar equal to $90°$.

Information about all the detected obstacles is collected in a single array of obstacles and is used for the vehicle orientation. At each step of the vehicle's motion, the array provides information about the obstacles that get into the "working area" shown in Fig. 3.9, if the following condition is satisfied

$$D_i < R,$$

where D_i is the distance to the *i*th obstacle and R the radius of the vehicle's working area.

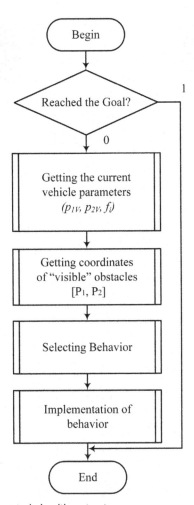

FIGURE 3.8 Vehicle control algorithm structure.

The supervision layer analyzes the current situation and selects a specific behavior for the next step of the vehicle's motion.

The behavior-selection priority is presented in Fig. 3.10 and is defined as follows:

$$\beta_1 > \beta_2 > \beta_3.$$

The selected behavior is executed giving the parameters for the vehicle.

The new coordinates for the vehicle's motion (p_{1V+1}, p_{2V+1}) are defined as follows:

$$p_{1V+1} = p_{1V} + V \cdot \cos(f_{i+1}) \cdot dt$$

$$p_{2V+1} = p_{2V} + V \cdot \sin(f_{i+1}) \cdot dt$$

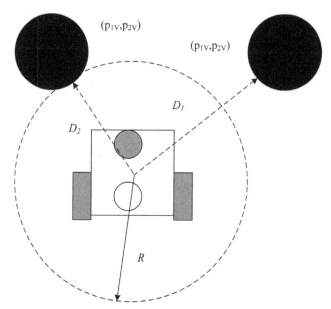

FIGURE 3.9 Vehicle's "working area."

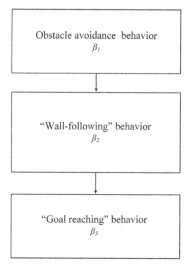

FIGURE 3.10 Behavior priorities.

where V is the vehicle's linear velocity; f_{i+1}, vehicle's orientation angle with respect to the goal; dt, path-planning time step.

This ends the current iteration and the algorithms goes to a new iteration to check the global condition "Reached the goal?".

3.5.2 Behavior Formation General Principle

According to the basic approach of Ref. [43] let us introduce each behavior as union of fuzzy rules in the following form:

$$R^{(l)}: \text{If } x_1 \text{ is equal to } A_1^l \text{ and...and } x_n \text{ is equal to } A_n^l,$$
$$\text{then } y \text{ is equal to } B^l, \tag{3.10}$$

where $l = 1, ..., m$, and m is the number of rules in a given fuzzy rule base; $x_1, ..., x_n$ are the input variables which are the sensor data of the vehicle; $A_1^l, ..., A_m^l$ are the input fuzzy sets, B^l is the output fuzzy set and y is the output variable.

An individual behavior is selected based on situation analysis. In this case, the information about the state of the environment is formed as a collection of sensor and radar data together with the information about the obstacles from the obstacle base created using the measurements results and analysis of the previous motion.

Analysis of the works [34,36,43] has shown that sector representation is convenient for obstacle detection. In the case considered, six sectors are used: *FL, FR, RU, RD, LU*, and *LD* (see Fig. 3.11).

The obstacles are clustered according to the algorithm presented in Fig. 3.12. It has the following steps.

Step 1. Obtaining obstacle coordinates as a combination of radar data and information from the obstacle base.

Step 2. Selection of obstacles falling into the vehicle's working area.

Step 3. Clusterization of obstacles into the groups according to sectors shown in Fig 3.12.

Step 4. Transferring data to the inputs of the corresponding software module.

Let us introduce a LV *Steering* to represent the vehicle's steering angle having the following term-set:

$$T(Steering) = \{R, \text{ Right}; RF, \text{ Right front}; F, \text{ Front}; LF, \text{ Left front}; L, \text{ Left}\}.$$

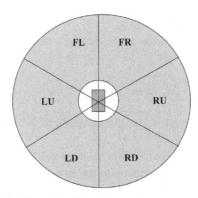

FIGURE 3.11 Vehicle position analysis sectors.

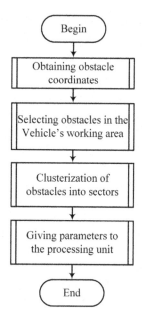

FIGURE 3.12 Algorithm of obstacles clusterization.

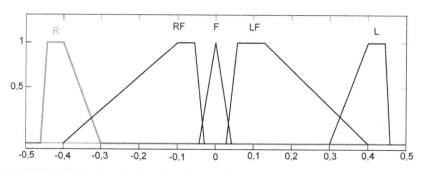

FIGURE 3.13 Membership functions for the term-set T(Steering).

The membership functions for the elements of the term-set of LV *T(Steering)* are presented in Fig. 3.13.

To describe the output values of the vehicle's velocity, an LV is introduced. It has the following term-set:

$$T(Velocity) = \{Z, \text{Zero}; \; SP, \text{Small positive}; \; P, \text{Positive}\}.$$

The membership functions for the elements of the term-set *T(Velocity)* is presented in Fig. 3.14.

The produced values of parameters *Steering* and *Velocity* specify the necessary motion of the vehicle in relation to the environmental state.

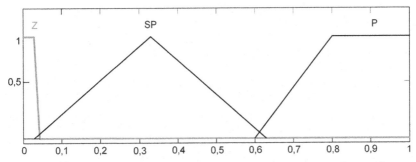

FIGURE 3.14 Membership functions for the elements of the term-set $T(Velocity)$.

Triangular and trapezoidal membership functions are selected for the sake of simplicity of their implementation in onboard computers of the vehicles.

Let's consider the principles of generation of independent behaviors.

3.5.3 Goal-Reaching Behavior

Goal-reaching behavior tends to drive the robot from a given initial position to a goal position, described in Ref. [43] in detail. This behavior drives the robot to left, right, or forward, depending on θ_{error}, the difference between the desired heading (the heading required to reach the goal) and the actual current heading D_{VG}. Fig. 3.15 illustrates "goal-reaching" behavior.

The inputs of the fuzzy controller implementing goal reaching receive the measured values of vehicle position coordinates. Knowing the goal coordinates, the distance to the goal D_{VG} is given by

$$D_{VG} = \sqrt{(p_{1G} - p_{1V})^2 + (p_{2G} - p_{2V})^2}. \tag{3.11}$$

and θ_{error} given by

$$\theta_{error} = \tan^{-1}\left(\frac{p_{2G} - p_{2V}}{p_{1G} - p_{1V}}\right) - f_i \tag{3.12}$$

where (p_{1V}, p_{2V}, f_i) is the vehicle position and heading, and (p_{1G}, p_{2G}) is the goal position.

LV used as controller inputs are:

- heading error θ_{error};
- distance to the goal D_{VG}.

Let us define the LV θ_{error} with the following term-set:

$$T(\theta_{error}) = \{N,\ Negative;\ SN,\ Small\ negative;\ Z,\ Zero;$$
$$SP,\ Small\ positive;\ P,\ Positive\}.$$

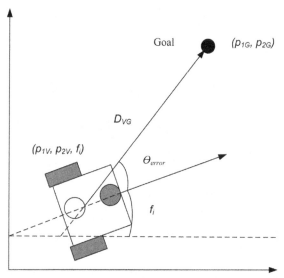

FIGURE 3.15 Goal-reaching behavior.

The membership functions of the elements of the term-set $T(\theta_{error})$ are presented in Fig. 3.16.

LV $T(D_{VG})$ has the following term-set

$$T(D_{VG}) = \{N, \text{ near}; \ S, \text{ small}; \ B, \text{ big}\}.$$

The membership functions of the elements of the term-set $T(D_{VG})$ are presented in Fig. 3.17.

The behavior "goal reaching" is expected to align the robot's heading with the direction to the goal minimizing the error value θ_{error}.

The vehicle's heading error lies in the interval [−180 degrees; 180 degrees], which eliminates definition ambiguity.

The algorithm executing goal-reaching behavior is a succession of actions presented in Fig. 3.18. It includes the following steps:

Step 1. Get current vehicle position coordinates.

Step 2. Calculate distance to the goal D_{VG}.

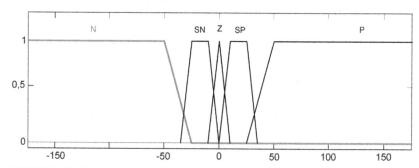

FIGURE 3.16 Membership functions of the elements of the $T(\theta_{error})$ term-set.

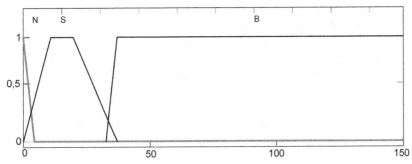

FIGURE 3.17 Membership functions of the elements of the $T(D_{VG})$ term-set.

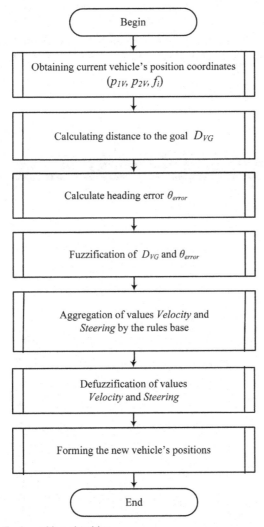

FIGURE 3.18 Goal-reaching algorithm.

Step 3. Calculate heading error θ_{error}, the difference between the desired heading and the actual current heading.

Step 4. Fuzzification of D_{VG} and θ_{error} values.

Step 5. Aggregation of output values by the rules base and generation of output variables values *Velocity* and *Steering*.

Step 6. Defuzzification of output variables *Velocity* and *Steering*.

Step 7. Forming the new vehicle's positions.

Based on the logical analysis of the vehicle motion principles and on the results of imitation modeling, a rules base presented in Table 3.1 was designed.

3.5.4 Obstacle-Avoidance Behavior

The obstacle-avoidance behavior tends to avoid collisions with obstacles in the vehicle's vicinity.

The inputs of the fuzzy controller receive the values of minimal distances to the obstacles for each of the sectors: *FL*, *FR*, *RU*, *RD*, *LU*, and *LD* (see Fig. 3.11).

Table 3.1 Control Rules Base for Goal-reaching Behavior

Rule Number	Input Variables		Output Variables	
1	θ_{error}	D_{VG}	*Steering*	*Velocity*
2	Z	N	F	Z
3	SN	N	RF	Z
4	N	N	R	Z
5	SP	N	LF	Z
6	P	N	L	Z
7	Z	S	F	P
8	SN	S	RF	P
9	N	S	R	SP
10	SP	S	LF	SP
11	P	S	L	SP
12	Z	B	F	P
13	SN	B	RF	P
14	N	B	R	SP
15	SP	B	LF	P
16	P	B	L	SP

Each of the sectors is presented by a corresponding LV. A term-set for the LV FrontLeft (*FL*) has the following form:

$$T(FL) = \{N, \text{ Near}; \ M, \text{ Medium}; \ F, \text{ Far}\}.$$

The term-sets for the other LVs are set analogously. The outputs are the LVs *Steering* and *Velocity*. The algorithm implementing the obstacle-avoidance behavior presented in Fig. 3.19 consists of the following steps:

Step 1. Getting the obstacles coordinates.

Step 2. Clusterization of obstacles into sectors. Determining distance to the obstacles and their orientation angle with respect to the vehicle.

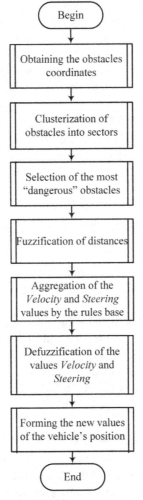

FIGURE 3.19 Obstacle-avoidance algorithm.

Step 3. Selection of the most "dangerous" obstacle based on the distance criterion.

Step 4. Fuzzification of the distance variable values for each sector.

Step 5. Aggregation of the output values by the rules base and generation of the correcting values of the output variables *Velocity* and *Steering*.

Step 6. Defuzzification of the output parameters *Velocity* and *Steering*.

Step 7. Forming the new values of the vehicle's position.

Based on the logical analysis of obstacle-avoidance principles and on the imitation modeling results, a control rules base was created for the obstacle-avoidance behavior. It is presented in Table 3.2.

The rules base for the obstacle-avoidance behavior differs from the one proposed in Ref. [43]. It includes 48 control rules instead of the original 20. It had to be modified because it was incomplete for the case of environments with obstacles of complicated shapes.

The number of used input LVs for obstacle-voidance behavior and the capacity of their term-set suggest a big amount of control rules (more than 700), which means the number of situations that can be processed by the controller is not full but their number is enough to pass rather complicated scenes.

The complexity of a fuzzy controller grows with the number of control rules used. This problem can be solved by designing a hierarchical fuzzy controller.

3.5.5 Wall-Following Behavior

The objective of the control of the wall-following behavior is to keep the vehicle at a close but safe distance from the wall and in line with wall.

The inputs of the fuzzy controller receive the values of minimal distances to the obstacles for each of the sectors: *FL*, *FR*, *RU*, *RD*, *LU*, and *LD*. LVs are defined analogously to the obstacle-avoidance behavior.

The outputs are the same LVs *Steering* and *Velocity*. The algorithm for wall-following behavior is the same as the one for obstacle avoidance, but has a different rules base.

Generally, this algorithm is activated if obstacles appear to the left or right of the vehicle at a medium distance in several sectors simultaneously. Fig. 3.20 presents such a situation.

This situation can be described by the following fuzzy rule.

If *RD* is *M* and *RU* is *M*, and *FR* is *VF* and *FL* is *VF*, and *LU* is *VF* and *LD* is *VF*, then *Steering* is *F* and *Velocity* is *SP*.

Based on the logical analysis of the wall-following principles and on the imitation modeling results, a control rules base was created for the wall-following behavior. It is presented in Table 3.3.

3.5.6 Behaviors Coordination Mechanism

For selection of behavior to achieve the best solution and form the output control signals *Steering* and *Velocity* for the actuators, it is necessary to account for the environmental conditions and priority of each behavior in the current situation.

Table 3.2 Control Rules Base for the Obstacle-Avoidance Behavior

Rule No.	Input Variables						Output Variables	
	RD	**RU**	**FR**	**FL**	**LU**	**LD**	**Steering**	**Velocity**
1	F	F	F	N	F	F	R	Z
2	F	F	F	M	F	F	R	SP
3	F	F	N	F	F	F	L	Z
4	F	F	M	F	F	F	L	SP
5	F	N	F	F	F	F	L	Z
6	F	M	F	F	F	F	L	SP
7	F	F	F	F	F	F	R	Z
8	F	F	F	F	M	F	RF	P
9	F	F	F	F	F	N	R	Z
10	N	F	F	F	F	F	L	Z
11	F	N	N	F	F	F	L	Z
12	F	M	M	F	F	F	L	SP
13	F	F	F	N	N	F	R	Z
14	F	F	F	M	M	F	R	SP
15	F	F	N	N	F	F	L	Z
16	F	F	M	M	F	F	L	SP
17	F	F	F	F	F	F	F	P
18	F	N	F	F	N	F	F	SP
19	F	N	F	M	N	F	F	SP
20	F	N	M	F	N	F	F	SP
21	F	F	N	N	F	F	LF	Z
22	F	F	N	N	N	F	RF	Z
23	F	N	N	N	F	F	LF	Z
24	N	N	N	F	F	F	LF	Z
25	F	F	F	N	N	N	RF	Z
26	N	N	M	F	F	F	LF	Z
27	F	F	F	M	N	N	RF	Z
28	N	N	N	N	F	F	L	Z
29	F	F	N	N	N	N	R	Z
30	F	N	N	N	N	F	L	Z
31	F	M	M	N	N	F	RF	Z
32	F	N	N	M	M	F	LF	Z

Continued

Table 3.2 Control Rules Base for the Obstacle-Avoidance Behavior continued

Rule No.	Input Variables						Output Variables	
	RD	RU	FR	FL	LU	LD	Steering	Velocity
33	F	M	M	N	N	M	RF	Z
34	M	N	N	M	M	F	LF	Z
35	F	F	N	N	N	M	RF	Z
36	M	N	N	N	F	F	LF	Z
37	F	F	F	N	N	N	RF	Z
38	N	N	N	F	F	F	LF	Z
39	F	F	N	N	N	N	RF	Z
40	N	N	N	N	F	F	LF	Z
41	N	M	M	N	N	N	RF	Z
42	N	M	M	N	N	M	RF	Z
43	N	N	N	M	M	N	LF	Z
44	M	M	N	M	M	N	LF	Z
45	F	F	M	M	N	N	RF	Z
46	F	F	M	N	N	N	RF	Z
47	F	F	N	N	N	N	RF	Z
48	N	N	N	M	F	F	LF	Z

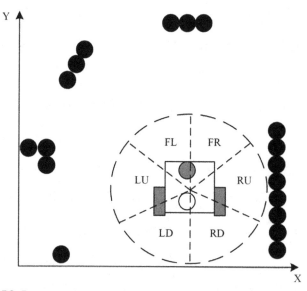

FIGURE 3.20 Presence of obstacle at the medium distance (*M*) at several sectors.

Table 3.3 Control Rules Base for the Wall-following Behavior

Rule No.	Input Variables						Output Variables	
	RD	RU	FR	FL	LU	LD	Steering	Velocity
1	M	M	F	F	VF	VF	F	SP
2	M	M	VF	VF	VF	VF	F	SP
3	M	M	M	F	VF	VF	F	SP
4	M	M	M	VF	VF	VF	F	SP
5	VF	VF	F	F	M	M	F	SP
6	VF	VF	VF	VF	M	M	F	SP
7	VF	VF	F	M	M	M	F	SP
8	VF	VF	VF	M	M	M	F	SP

Based on the results presented in Ref. [36,43], architecture for implementation of this approach is introduced (Fig. 3.21).

The outputs of the control system, *Steering*, S_V, and *Velocity*, V_V, are calculated using the values of S_1, V_1, S_2, V_2, S_3, and V_3 obtained for each individual behavior. This requires an analysis of distance to the goal D_{VG} and heading error θ_{error} together with values of distances to the obstacles in the vehicle's working area for each of the sectors.

Unlike the supervisor used in the work [43], the number of logical rules used to determine the behavior activation priority is reduced from 13 to 7. The obstacle-avoidance behavior is activated by default. Later on, the system checks if other behaviors should be activated.

To avoid conflict of *obstacle-avoidance* and *goal-reaching* behaviors, the last one is blocked when heading error exceeds the value of $65°$ ($|\theta_{error}| > 65°$).

3.5.7 Vehicle Control Simulation Results

To verify the validity of the proposed scheme, some typical cases are simulated in which a vehicle is to move from a given current position to a desired goal position in various unknown environments. In all cases, the vehicle is able to navigate its way toward the goal while avoiding obstacles successfully.

Fig. 3.22 shows a scene with a simple obstacles arrangement. The vehicle moves from the initial point (0,0) to the goal (20,20) using the goal-reaching behavior. When the vehicle approaches an obstacle, it changes its behavior to obstacle avoidance. The modeling results demonstrate successful reaching of the goal.

An additional example of avoiding simple obstacles is presented in Fig. 3.23 (L-obstacle) and Fig. 3.24 (I-obstacle). The modeling results show that the vehicle control system solves the goal-reaching task successfully.

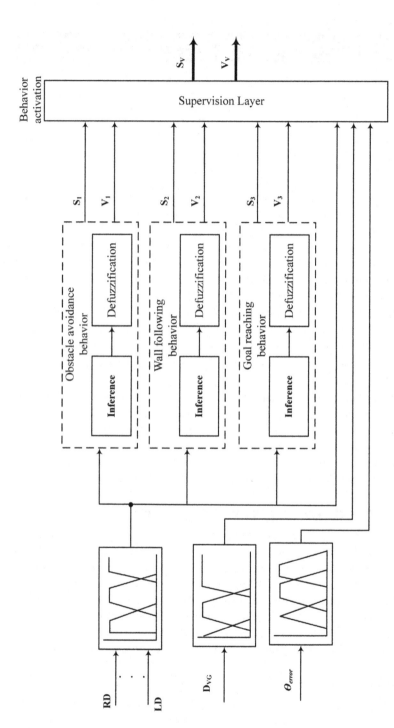

FIGURE 3.21 Control system architecture. S_i and V_i, steering and velocity for each control strategy, where $i = 1, ..., 4$; S_V and V_V, output values of steering and velocity, respectively, supplied to the vehicle's actuators.

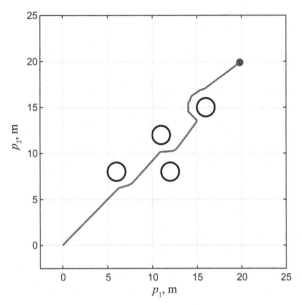

FIGURE 3.22 Simple obstacle arrangement scene.

Fig. 3.25 shows avoiding sparsely located obstacles, while in Fig. 3.26 they are positioned closely together. The arrangements of obstacles in Figs. 3.25 and 3.26 are different, but in both the cases the vehicle successfully moves from the

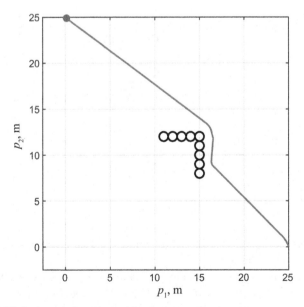

FIGURE 3.23 L-shaped obstacle-avoidance scene.

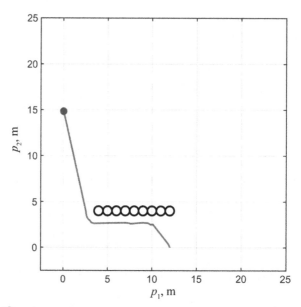

FIGURE 3.24 I-shaped obstacle-avoidance scene.

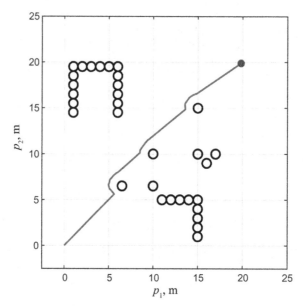

FIGURE 3.25 Avoiding sparse obstacles.

initial point (0,0) to the goal (20,20) blending the goal-reaching behavior with obstacle-avoidance behavior.

Fig. 3.27 illustrates the situation of avoiding L-shaped obstacle moving from the initial point (0,0) to the goal (20,20). The modeling has demonstrated effective task completion by the control system.

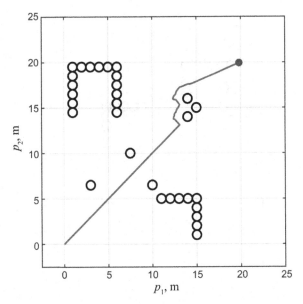

FIGURE 3.26 Avoiding closely located obstacles.

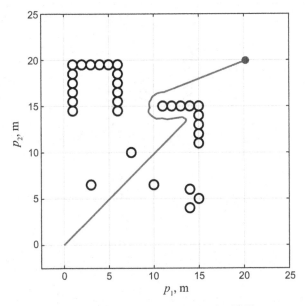

FIGURE 3.27 Avoiding L-shaped obstacle starting at point (0,0).

Figs. 3.28–3.30 illustrate the situations of L-shaped obstacle avoidance for various initial positions. The vehicle's control system "sees" and analyzes the environment differently causing generation of different paths for the vehicle. However, the control system successfully solves the goal-reaching task.

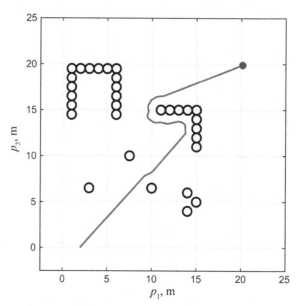

FIGURE 3.28 Avoiding L-shaped obstacle starting at point (2,0).

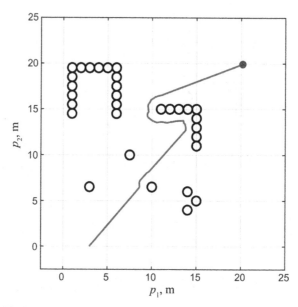

FIGURE 3.29 Avoiding L-shaped obstacle starting at point (3,0).

The examples of escape from a Π-shaped obstacle are presented in Figs. 3.31 and 3.32.

As presented in Fig. 3.31, the vehicle turns around and performs wall-following, keeping a constant orientation angle. The wall-following behavior is deactivated after the obstacle is passed.

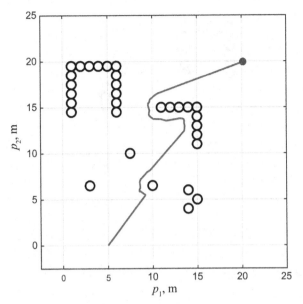

FIGURE 3.30 Avoiding L-shaped obstacle starting at point (5,0).

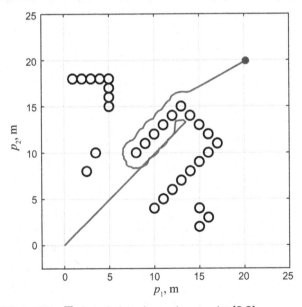

FIGURE 3.31 Avoiding Π-shaped obstacle starting at point (0,0).

As presented in Fig. 3.32, the vehicle uses the wall-following behavior and gets out of the Π-obstacle. The control system solves the task successfully.

The values of performance criteria described earlier for the performed modeling are presented in Table 3.4.

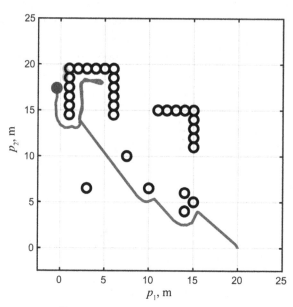

FIGURE 3.32 Avoiding Π-shaped obstacle starting at point (20,0).

Table 3.4 Fuzzy Control System Performance Criteria

Performance Criterion	Fuzzy Planner
Safety criterion, S_m	0.44829
	0.59352
	0.35418
	0.74528
	0.33253
Path length, P_L	36.1179
	35.7002
	35.8250
	42.9883
	54.1726
Task completion time, t_m	72.2358
	71.4004
	71.65
	85.9766
	108.3452
Mission success coefficient, F	1
	1
	1
	1
	1

The simulation results demonstrate effectiveness of the presented approach that uses a fuzzy controller and verbal description of input parameters.

3.6 SUMMARY

A lot of control tasks for vehicles and their groups are to be solved for unknown dynamic environment with partially defined models. Solution of control tasks under the conditions of uncertainty requires taking decisions directed to the search of control actions ensuring optimal values of the set criteria functions determining effectiveness of the control system's functioning [58–72].

Decision making under the conditions of uncertainty is essentially facilitated by the application of fuzzy sets and fuzzy logic. It allows us to reduce uncertainty and compensate for information incompleteness. This approach is based on the human ability to process information at the perception level. The fuzzy-logic rules give a formal methodology for linguistic rules ensuing form reasoning and decision making based on fuzzy and inexact information.

In vehicle control tasks, especially ones that function in critical modes, control is considered as a process of decision making under the conditions of uncertainty. This chapter addresses the peculiarities of vehicle control under the conditions of uncertainty using fuzzy schemes. A classification of uncertainty types is presented in application to environment description for vehicle control tasks. A common fuzzy control sequence was considered.

Formalization of spatial relations between the vehicle and the environment was performed using extensional and intentional relations. The vehicle's behavior was formalized based on the classification models and calculation of truth degree of fuzzy production rules in the situational model.

The analysis of existing vehicle motion control methods is performed and a method of context-based behavior is proposed. This method splits the overall behavior into small independent behaviors that focus on execution of a specific subtask.

Several specific behaviors were introduced such as "goal reaching," "obstacle avoidance," "wall-following." A method of fusion of behaviors is introduced. Simulation results for the vehicle functioning in different obstacle location scenes are presented. The results confirming the effectiveness of the proposed method were obtained for the vehicle path planning under the conditions of environmental uncertainty.

REFERENCES

[1] V. Kh Pshikhopov, M. Yu Medvedev, Estimation and Control in Complex Dynamic Systems, Fizmatlit, Moscow, 2009.
[2] V.A. Bessekerskiy, E.P. Popov, Theory of the Automated Control Systems, fourth ed., Profession Publishing House, Saint Petersburg, 2003.
[3] A.R. Gajduk, Continuous and Discrete Dynamic Systems, second ed., Educational Methodological and Publishing Center, Moscow, 2004 (Educational Literature).
[4] L.S. Bernstein, V.P. Karelin, A.N., Celyh, Models and Methods of Decision Making in Integrated Intelligent Systems, Rostov University, Rostov-on-Don, 1999.

[5] S.I. Rodzin, Decision-making Theory: Lecture and Practical Work: Tutorial, SFEDU, Taganrog, 2010.

[6] V. Kh Pshikhopov, N. Sergeev, M. Medvedev, A. Kulchenko, The design of helicopter autopilot, SAE Technical Papers (2012).

[7] V. Kh Pshikhopov, V.A. Krukhmalev, M. Yu Medvedev, et al., Adaptive control system design for robotic aircrafts, IEEE Latin American Robotics Symposium (2013) 67−70.

[8] S.E. Bublej, Yu A. Zargaryan, System approach for control tasks solving in the conditions of indeterminacy, in: Proceedings of the Congress for Intelligent Systems and Information Technologies AIS-IT, vol. 3, Fizmatlit, Moscow, 2010, pp. 273−276.

[9] V.I. Finaev, Modeling during Information-Control Systems Design, TSURE, Taganrog, 2002.

[10] L.A. Zade, Fuzzy Sets and Their Application in Image Identification and Cluster Analysis, Classification and a Cluster, Mir Publishers, Moscow, 1980, pp. 208−247.

[11] A. Kofman, Introduction to the Fuzzy Sets Theory, Radio and Communications, Moscow, 1982.

[12] A.N. Averkin, I.Z. Batyrshin, A.F. Blinshun, Fuzzy Sets in Control and Artificial Intellect Models, Nauka, Moscow, 1986.

[13] T.L. Saaty, Measuring the fuzziness of sets, Journal of Cybernetics 4 (1974) 149−194.

[14] D. Du Bois, A. Prad, in: V.B. Tarasova (Ed.), Theory of Possibilities, Radio and Communications, Moscow, 1990 (translation from French).

[15] A.N. Borisov, O.A. Krumberg, I.P. Fedorov, Decision Making Based on Fuzzy Models: Examples of Application, Zinatne, Riga, 1990.

[16] D.A. Pospelov, Situational Control: Theory and Practice, Nauka, Moscow, 1988 (Chief Editor of Physico-mathematical Literature).

[17] V.I. Finaev, Decision-Making Models, TSURE, Taganrog, 2005.

[18] V.I. Ivanenko, V.A. Labkovsky, Indeterminacy Problem in Decision-making Tasks, Naukova Dumka, Kiev, 1990.

[19] E. Yu Kandrashina, D.A. Litvintseva, L.V. Pospelov, Time and Space Knowledge Representation in Intelligent Systems, Nauka, Moscow, 1989.

[20] A.S. Yushchenko, Methods of fuzzy logic in the control of mobile manipulation robots, Vestnik MGTU, in: N. Uh Bauman (Ed.), Ser. Priborostroenie, MGTU Publishing Office, Moscow, 2012, pp. 29−43.

[21] E.H. Mamdani, Advances in the linguistic synthesis of fuzzy controllers, International Journal of Man-Machine Studies 8 (1976) 669−678.

[22] E.H. Mamdani, Application of fuzzy logic to approximate reasoning using linguistic synthesis, IEEE Transactions on Computers 26 (12) (1977) 1182−1191.

[23] A.S. Yushchenko, Robot control using fuzzy logics: state and problems, News of the Artificial Intellect 1 (2006) 119−130.

[24] L.S. Bershtejn, A.V. Bozhenyuk, Fuzzy Decision-making Models: Deduction, Induction, Analogy, TSURE Publishing Office, Taganrog, 2001.

[25] L.S. Bershtejn, V.I. Finaev, Adaptive Control with Fuzzy Strategies, Rostov University, Rostov-on-Don, 1993.

[26] A.N. Melihov, L.S. Bershtejn, S. Ya Korovin, Situational Advising Systems with Fuzzy Logics, Nauka, Moscow, 1990.

[27] T. Terano, K. Asan, M. Ougeno (Eds.), Applied Fuzzy Systems: Translation from Japanese, Mir Publishers, Moscow, 1993.

[28] L.S. Bershtejn, A.V. Bozhenyuk, Fuzzy Logical Judgment Based on Definition of the Fuzzy Rule Modus Ponens Truth Definition, Methods and Systems of Decision Making, Systems based on the knowledge, RPI, Riga, 1989, pp. 74−80.

[29] L.A. Zaden, Fuzzy sets, Information and Control 8 (1965) 338−353.

[30] V.V. Vojnov, A.S. Yushchenko, Control of a micro robot for medical purpose using fuzzy finite-state automaton and situation control methods, extreme robotics. Nano-, micro- and macro robots, in: Proceedings of the International Conference, Polytechnica-service, Gelendzhik, Saint Petersburg, 2009, pp. 115−116.

[31] A.S. Yushchenko, Intelligent planning in robots' activity, Mechatronica 3 (2005) 5−18.

[32] A.A. Zdanov, Autonomous Artificial Intellect, BINOM, Moscow, 2008.

[33] D.Yu Brovkina, T.A. Prihodko, Development of obstacle avoidance algorithm for mobile controlled robot, in: Scientific Conference Information Control Systems and Computer Monitoring (ICS CM 2013), 2013, pp. 415—421.

[34] T.S. Hong, D. Nakhaeinia, B. Karasfi, Application of Fuzzy Logic in Mobile Robot Navigation, University Putra Malaysia, Malaysia, Fuzzy Logic Controls, Concepts, Theories and Applications Source, InTech, 2012, pp. 22—36.

[35] A.S. Al Yahmedi, E. El-Tahir, T. Pervez, Behavior based control of a robotic based navigation aid for the blind, in: Control & Applications Conference "CA-2009", Cambridge, UK, July 13—15, 2009.

[36] A.S. Al Yahmedi, A.M. Fatmi, Fuzzy logic based navigation of mobile robots, Sultan Qaboos University, Oman, International Journal of Advanced Robotic Systems, 2011, pp. 287—310.

[37] R.C. Arkin, T. Balch, AuRA: principles and practice in review, Journal of Experimental and Theoretical Artificial Intelligence (JETAI) 9 (2/3) (1997) 175—188.

[38] R.C. Arkin, Motor schema-based mobile robot navigation, International Journal of Robotic Research 8 (1989) 92—112.

[39] R.C. Arkin, Towards Cosmopolitan Robots: Intelligent Navigation in Extended Manmade Environments (Ph.D. thesis), University of Massachusetts, Department of Computer and Information Science, 1987.

[40] R.A. Brooks, A robot that walks; emergent behavior from a carefully evolved network, in: IEEE International Conference on Robotics and Automation, Scottsdale, AZ, 1989, pp. 292—296.

[41] R.A. Brooks, A robust layered control system for a mobile robot, IEEE Journal of Robotics and Automation 2 (1) (1986) 14—23.

[42] T. Ching-Chih, C. Chin-Cheng, C. Cheng-Kain, L. Yi Yu, Behavior-based navigation using heuristic fuzzy kohonen clustering network for mobile service robots, International Journal of Fuzzy Systems 12 (1) (2010) 25—32.

[43] A. Fatmi, A.S. Al Yahmedi, L. Khriji, N.A. Masmoudi, Fuzzy logic based navigation of a mobile robot, World Academy of Science, Engineering and Technology 22 (2006) 169—174.

[44] R.G. Huq, K.I. Mann, R.G. Gosine, Mobile robot navigation using motor schema and fuzzy context dependent behavior modulation, Applied Soft Computing 8 (2008) 422—436.

[45] D. Langer, J.K. Rosenblatt, M.A. Hebert, Behavior-based system for off-road navigation, IEEE Journal of Robotics and Automation 10 (6) (1994) 776—782.

[46] P. Maes, How to do the right thing, Connection Science Journal 1 (1990) (special issue on Hybrid Systems).

[47] M.J. Mataric, Behavior-based control: examples from navigation, learning, and group behavior, Journal of Experimental and Theoretical Artificial Intelligence 9 (2/3) (1997) 323—336 (special issue on Software Architectures for Physical Agents).

[48] J. Rosenblatt, DAMN: A Distributed Architecture for Mobile Navigation (Ph.D. dissertation), Carnegie Mellon University, Robotics Institute, Pittsburgh, PA, 1995. Technical Report CMU-RITR-97—01.

[49] J. Rosenblatt, D.W. Payton, A fine-grained alternative to the subsumption architecture for mobile robot control, in: Proceedings of the IEEE/INNS International Joint Conference on Neural Networks, Washington, DC, vol. 2, June 1989, pp. 317—324.

[50] A. Saffiotti, The uses of fuzzy logic for autonomous robot navigation: a catalogue raisonné, Soft Computing Research Journal 1 (4) (1997) 180—197.

[51] P. Althaus, H.I. Christensen, Behavior coordination for navigation in office environment, in: Proceedings of 2002 IEEE/RSJ International Conference on Intelligent Robots and Systems, 2002, pp. 2298—2304.

[52] E. Aguirc, A. Gonzalez, Fuzzy behaviors for mobile robot navigation: design, coordination and fusion, International Journal of Approximate Reasoning 25 (2000) 255—289.

[53] M.F. Selekwa, D.D. Dunlap, E.G. Collins Jr., Implementation of multi-valued fuzzy behavior control for robot navigation in cluttered environments, in: Proceedings of the IEEE International Conference on Robotics and Automation, Barcelona, Spain, 2005, pp. 3699—3706.

[54] H. Seraji, A. Howard, Behavior-based robot navigation on challenging terrain: a fuzzy logic approach, IEEE Transactions on Robotics and Automation 18 (3) (2002) 308–321.

[55] S.X. Yang, H. Li, M.Q.H. Meng, P.X. Liu, An embedded fuzzy controller for a behavior-based mobile robot with guaranteed performance, IEEE Transactions on Fuzzy Systems 12 (4) (2004) 436–446.

[56] S.X. Yang, M. Moallem, R.V. Patel, A layered goal-oriented fuzzy motion planning strategy for mobile robot navigation, IEEE Transactions on Systems, Man, and Cybernetics. Part B: Cybernetics 35 (6) (2005) 1214–1224.

[57] E. Tunstel, T. Lippincott, M. Jamshidi, Behavior hierarchy for autonomous mobile robots: fuzzy-behavior modulation and evolution, International Journal of Intelligent Automation and Soft Computing 3 (1) (1997) 37–49 (special issue: Autonomous Control Engineering at NASA ACE Center).

[58] V.I. Finaev, Yu A. Zargaryan, Dynamic process parameters optimization method under conditions of incomplete data, Bulletin of the Rostov State University of the Railroads 3 (2011) 74–78, 524.

[59] A. Chinchuluun, P.M. Pardalos, R. Enkhbat, I. Tseveen, Optimization and Optimal Control: Theory and Applications, Springer-Verlag, New York, 2010, p. 524.

[60] V.I. Finaev, Yu Zargaryan, Multicriterion decision making in case of incomplete source data, World Applied Sciences Journal 23 (9) (2013) 1253–1261.

[61] G. Eichfelder, Adaptive Scalarization Methods in Multiobjective Optimization, Springer, 2008, p. 256.

[62] R. Neydorf, V. Krukhmalev, N. Kudinov, V. Kh Pshikhopov, Methods of statistical processing of meteorological data for the tasks of trajectory planning of MAAT feeders, SAE Technical Papers (2013).

[63] M. Panos, A. Pardalos Migdalas, L. Pitsoulis, Pareto Optimality, Game Theory and Equilibria, first ed., Springer-Verlag, New York, 2008, p. 888.

[64] P.M. Pardalos, V.A. Yatsenko, Optimization and Control of Bilinear Systems: Theory, Algorithms, and Applications, Springer, 2010, p. 396.

[65] V. Kh Pshikhopov, V. Krukhmalev, M. Medvedev, R. Neydorf, Estimation of energy potential for control of feeder of novel cruise/feeder MAAT, SAE Technical Papers (2012).

[66] V. Kh Pshikhopov, M. Medvedev, V. Kostjukov, et al., Airship autopilot design, SAE Technical Paper (2011).

[67] V. Kh Pshikhopov, M. Medvedev, R. Neydorf, et al., Impact of the feeder aerodynamics characteristics on the power of control actions in steady and transient regimes, SAE Technical Papers (2012).

[68] V. Kh Pshikhopov, M. Yu Medvedev, B.V. Gurenko, Homing and docking autopilot design for autonomous underwater vehicle, Applied Mechanics and Materials (2014) 490–491.

[69] V. Kh Pshikhopov, A.S. Ali, Hybrid motion control of a mobile robot in dynamic environments, in: International Conference on Mechatronics (ICM 2011), 2011, pp. 540–545.

[70] V. Kh Pshikhopov, M. Yu Medvedev, Robust control of nonlinear dynamic systems, in: IEEE ANDESCON Conference Proceedings, ANDESCON, 2010.

[71] V. Kh Pshikhopov, M. Yu Medvedev, A.R. Gaiduk, B.V. Gurenko, Control system design for autonomous underwater vehicle, in: Latin American Robotics Symposium, 2013, pp. 77–82.

[72] T.J. Ross, Fuzzy Logic with Engineering Applications, Wiley–Blackwell (an imprint of John Wiley & Sons Ltd.), 2012, p. 606.

Chapter Four

Genetic Algorithms Path Planning

V. Krukhmalev[1], V. Pshikhopov[2]

[1]RoboCV, Ltd., Moscow, Russia; [2]Southern Federal University, Taganrog, Russia

4.1 GENERALIZED PLANNING ALGORITHM

Genetic algorithms find many applications in solution of path planning problems for complicated environments with stationary obstacles. Genetic algorithms can be used even for mobile or unknown obstacles, given that we have sufficient computational power. This chapter proposes the structural and algorithmic solutions for the vehicle's path planning problem using genetic algorithms (GAs).

There are various methods and approaches to path planning using the graph theory [1–3]. Some of them propose path planning for aviation [4–11] using GAs [12]. The work [13] uses classical graph optimization methods for a ground-based vehicle path search. The proposed approach solves the planning task using a graph. An orientated graph serves as an environment model while GAs work as a tool for an optimal graph-searching for the positions of the robot and surrounding obstacles.

The algorithms proposed in this chapter assume the following succession of actions:

- Getting the current vehicle and goal coordinates;
- representing the entire functioning space in the form of an oriented graph where each vertex represents a possible discrete spatial position of the robot (with a set sampling), and edges show the possibilities of transitions between these positions;
- getting information about the obstacles;
- vehicle's initialization in the graph space and identification of the goal point in the graph space; changing the graph according to the obstacles arrangement;
- in the loop, until the goal is reached (goal vertex), a search for an optimal (shortest) graph is performed using GA methods by rebuilding the graph according to robot's position and visible obstacles; and
- transfer of path parameters from the planner to position-path controller.

Path Planning for Vehicles Operating in Uncertain 2D Environments. http://dx.doi.org/10.1016/B978-0-12-812305-8.00004-1

The further research is performed on the assumption that the point "getting the current vehicle and goal coordinates" is passed.

4.2 GRAPH FORMATION

4.2.1 General Concepts

Graph theory and graph-based algorithms find most of their applications in programming. They are preferred in discrete mathematics because they provide a convenient language for model description [14−25].

A graph is a model of a system of objects connected by certain relations. A graph $G(V,E)$ is an aggregate of two sets comprising a nonempty set V (a set of vertices) and a set of edges E, which are two-element subsets of V [23]:

$$G(V,E) = <V,E>, \quad V \neq 0 \qquad E \subset 2^V \text{ and } \forall e \in E(|e| = 2)$$

where V is a set of vertices and E a set of edges.

Each set E of two-element subsets of set V defines a symmetric binary relation on the set V. So we can assume,

$$E \subset V \times V, \quad E = E^{-1}$$

and treat an edge both as a set $\{e_1, e_2\}$ and as a pair (e_1, e_2).

The number of vertices of graph G can be denoted by p and the number of edges by q. Then, the following relation holds:

$$p = p(G) = |V|, \quad q = q(G) = |E|.$$

In the object system model, the objects correspond to graph vertices and the links among objects are the edges (arcs) of the graph. The graph is represented as a multitude of points (vertices) connected by lines (edges or arcs in the oriented graph).

Mathematically, a graph G is set as an aggregate of vertices and edges $G(V,E)$, where $V = \{v_1, v_2, ..., v_n\}$, set of vertices ($n$ is the number of vertices); and $E = \{e_1, e_2, ..., e_m\}$, set of edges ($m$ is the number of edges that can differ from the number of vertices).

The edges can be represented as $e(v_1, v_2)$ being an edge between vertices v_1 and v_2, or even as a pair (v_1, v_2). An example of a graph is presented in Fig. 4.1.

Usually graph is represented as: vertices by points and edges by lines. Fig. 4.2 shows an example of a diagram of a graph having four vertices and five edges.

In this graph, the vertices v_1 and v_2, v_2 and v_3, v_3 and v_4, and v_2 and v_4 are adjacent and vertices v_1 and v_3 are nonadjacent. Adjacent edges are: e_1 and e_2, e_2 and e_3, e_3 and e_4, e_4 and e_1, e_1 and e_5, e_2 and e_5, e_3 and e_5, and e_4 and e_5. Nonadjacent edges are: e_1 and e_3, and e_2 and e_4.

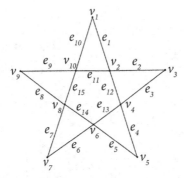

FIGURE 4.1 Example of a graph.

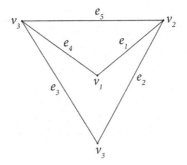

FIGURE 4.2 Diagram of a graph.

4.2.2 Algorithm of Graph Formation for the Planning Task

When solving a path-finding problem, the graph vertices represent a vehicle's all possible positions in space with set dimensions range and sampling rate of these positions in space that can also be a phase space [2]. Obviously, the practical application of the proposed approaches requires rather high sampling rates for the spatial positions, which can lead to reduction of computational efficiency for such a graph.

The graph edges between the vertices correspond to vehicle's motion parameters. The main idea of the proposed approach is to create a graph in such a way that the obstacles detected by the vehicle's sensor system could be represented by the vertices with no connected edges.

During the vehicle's motion at each planning step the following data is used for graph creation and rebuilding:
- Sampling rate;
- vehicle's initial position at the moment of control system initialization;
- vehicle's motion goal; and
- obstacles coordinates.

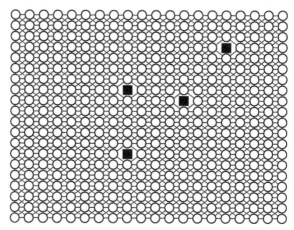

FIGURE 4.3 Modeling scene graph.

Vehicle's initial position becomes the center of the global coordinate system—the zero point. The sampling intervals for vertical and horizontal axes are considered equal to 1.

This data is used in the following way. A graph is created where each vertex is connected to all the neighboring vertices. The weight of each edge equals to one. Each vertex is named according to its spatial position. In this case,

$$N(x, y),$$

where $x = \overline{1, k}$ is vehicle's position on abscissa (in meters), $y = \overline{1, m}$, vehicle's position on ordinate (in meters); and k, m, maximal number of graph vertices for both axes.

The information about the obstacles is obtained from the vehicle's sensory subsystem. Vehicle's coordinates (x_p, y_p) are positioned at the graph vertices $N_p(x_p, y_p)$. The vertices are marked as obstacles and the edges around such vertices are deleted. An example of such a graph is presented in Fig. 4.3.

The optimal path is searched using GAs in the space of this created graph.

4.3 DEVELOPMENT OF GENETIC ALGORITHMS FOR PLANNING

4.3.1 General Notions of Genetic Algorithms Theory

Currently, there are many definitions of GAs [26—31]. We use the most common one in this chapter.

A GA is a search heuristic that mimics the process of natural selection using random selection, combination and variation of parameters. This heuristic used to generate useful solutions for optimization and modeling problems. Genetic algorithms belong to the larger class of evolutionary algorithms, which generate solutions to optimization problems using techniques inspired by natural evolution, such as inheritance, mutation, selection, and crossover. Genetic algorithms give major stress on using crossover operator that performs a recombination operation between the candidate solutions—an operation analogous to the natural crossover.

A flow chart of a GA searching for an optimal solution is presented in Fig. 4.4.

Before the first step, an initial population is generated. If it happens to be totally uncompetitive, the GA will most likely transform it into a viable population. Hence, there is no need to form mostly fit individuals at the first step. It is sufficient for them to comply with the individuals' format and be suitable for calculation of the fitness function. A result of the first step is a population H consisting of N individuals.

Crossover in GAs implies that in order to produce descendants, there must be several (at least two) parents.

Crossover is defined differently for different algorithms. It depends on data representation. The main requirement is that the descendant or descendants should be able to inherit traits of both the parents mixing them in a certain way.

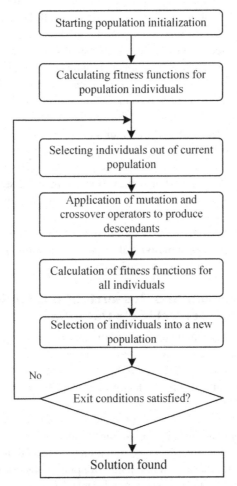

FIGURE 4.4 Modeling scene graph illustration.

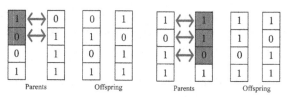

FIGURE 4.5 Crossover of individuals.

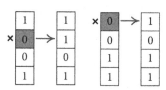

FIGURE 4.6 Mutations of individuals.

One of the main drawbacks of the early GAs is the small diversity of the individuals. A single, locally optimal genotype is formed rather quickly. All the other elements of the population are lost in selection and all the population becomes filled by the copies of that individual. During research, various ways of fighting such undesirable effects were developed. One such way is to perform a crossover not for mostly fit individuals but for all the individuals as shown in Fig 4.5.

In addition to the crossover operation, a mutation operation [27] is also applied. Mutations are treated the same way as multiplication. A certain group of mutants, m, is considered as a GA parameter. At the mutation step, a group of m_N individuals is selected. Then each of these individuals is modified according to the certain mutation operations. The mutation process is demonstrated in Fig. 4.6.

At the selection stage, some part of the population is selected to survive at the current evolution stage. At present, various methods of selection are developed and proved. Survival probability of an individual h should depend on the value of the fitness function fitness(h). The part of survived individuals s is usually a parameter of the GA and is set in advance. As a result of selection, s_N individuals are left out of N individuals of the population H to form the final population H'. All the other individuals perish.

4.3.2 Development and Research of a Genetic Algorithm for Vehicle's Path Planning

According to the classical theory of evolutionary computations for the considered problem, it is necessary to define the following steps:
1. Formalize the problem to be solved.
2. Solve the problem of representation of a path, being an optimal directed graph in the form of an individual.
3. Develop an algorithm estimating the individual's fitness (graph weight).
4. Develop a population initialization algorithm.
5. Select the operators of crossover, mutation, and selection.
6. Develop an exit criterion for automatic stopping of the optimization process indicating that an optimal solution is found.

A task of GAs is the minimization of the following function :

$$F = \sum_{i=1}^{n} S_i \qquad (4.1)$$

where S_i is an elementary transition along the graph edges between the nearest vertices, n, total number of transitions between the vertices, i.e., the path length in the graph domain.

Based on the fact that an individual should describe consecutive motions of the vehicle from the initial point to the final one, it follows that each gene of an individual should describe an elementary motion between the graph vertices connected by edges. Taking into account the 2D character of the graph and ordering of its nodes in the coordinates grid as shown in Fig. 4.3, each individual's gene is a pair of variables $[\Delta x_i, \Delta y_i]$ describing a shift from the previous vertex. Each of the variables from $[\Delta x_i, \Delta y_i]$ takes the value of "1", "-1," or "0". Both components cannot take zero values simultaneously. The shift direction is set in the classical Cartesian system of coordinates as shown in Fig. 4.7.

Thus, an individual formed out of the genes mentioned above is described by the following expression:

$$S = [\Delta x_i, \Delta y_i],$$

where $i = \overline{1, n}$, n is the maximal length of an individual.

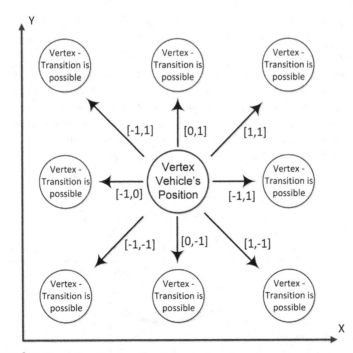

FIGURE 4.7 Transition directions in relation to the gene value.

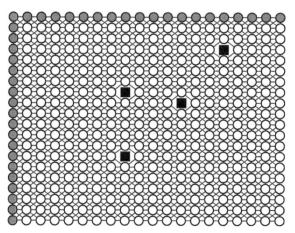

FIGURE 4.8 Path in graph with a length $n = 40$.

The initialization procedure for the starting population is rather straightforward. It uses a random selection out of three numbers "1", "0," and "−1". An important point is the selection of individual's dimension (length) n that should allow the individual to reach the most distant points on the graph. In the worst scenario, the optimal graph will be constructed at the edge of the initial graph with a certain margin for the possible obstacles located along this edge as shown in Fig. 4.8. In our case, its length is equal to

$$n = (X_G + Y_G) \times 1.25.$$

Since an individual is represented as an array of limited numbers, the most suitable type of mutation is the gene position change [28] and a two-point crossover type [29]. Universal tournament was used for a selection type [28].

Mutation is performed as follows. An arbitrary brake index r is selected. Then the two parts change places. The mutation scheme is presented in Fig. 4.9.

Two-point crossover is illustrated in Fig. 4.10. The two indices are selected in each individual. Then the two individuals exchange the selected gene sections.

The tournament selection marks out the mostly fit individuals a certain number of times out of an arbitrary population sample.

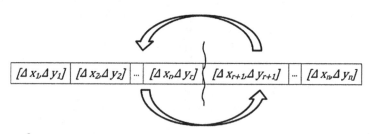

FIGURE 4.9 Gene position change.

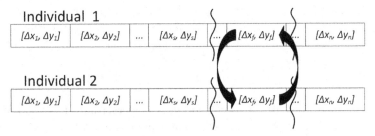

FIGURE 4.10 Two-point crossover.

In order to determine the stop criterion, it is necessary to proceed to the GAs testing.

In the modern theory of evolutionary calculations, there is a variety of approaches for implementation of GAs and strategies of using the genetic operators in the evolution process.

Let us mark out two algorithms of population variation. Population variation is defined as generation of an offspring (a new population) from a previous generation using mutation and crossover. The first algorithm illustrated in Fig. 4.11 uses a joint consecutive application of mutation with a probability of P_{mut} and crossover with a probability of P_{cx} for a selected individual.

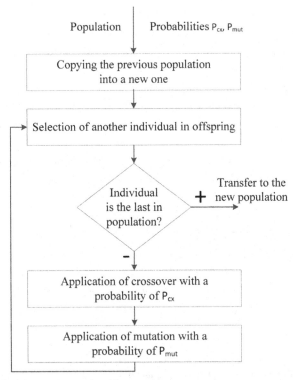

FIGURE 4.11 Joint variation algorithm.

The second algorithm presented in Fig. 4.12 uses either mutation or crossover for a selected individual with probabilities P_{cx} and P_{mut}, respectively.

Let us highlight the three types of GAs implementation using the described algorithms of population variation. The first fundamental approach [32] uses the algorithms of joint variation of generations. Its flowchart is presented in Fig. 4.13.

After we set the types of genetic operators, numeric parameters of evolution and initialization of population, the main evolutionary process starts up and

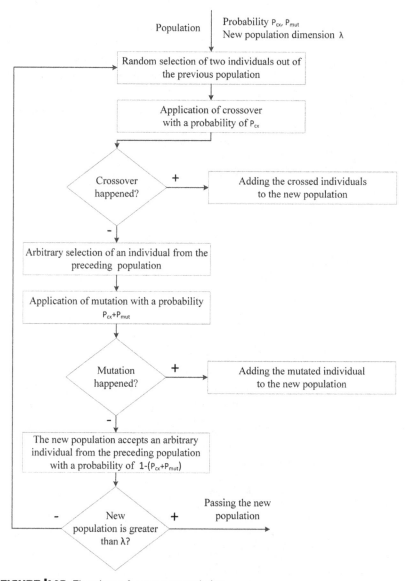

FIGURE 4.12 Flowchart of a separate variation.

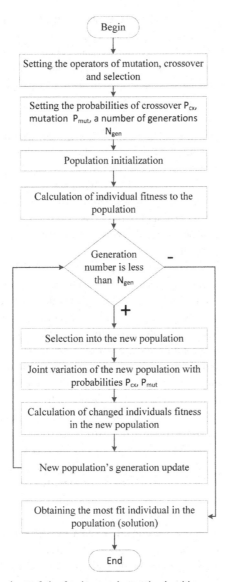

FIGURE 4.13 Flowchart of the fundamental genetic algorithm.

continues for a set number of generations. It implements the algorithm of joint variation of generations after a selection and update of population. The final solution is the best individuals of the final generation.

Two alternative approaches [32] have a tunable number of offspring λ, and a simultaneous tuning of the number μ of the individuals in the new generation selected out of the offspring and individuals of the preceding generation. They both use an algorithm of population's separate variation.

In the second approach, the offspring population having the size of λ is obtained by a crossover, mutation, or reproduction (cloning) with probabilities

P_{cx}, P_{mut} and $1 - (\mu + \lambda)$, respectively as shown in Fig. 4.12. The new generation is formed by selection of the offspring population and the preceding generation. Hereinafter, this type of algorithm is called as $(\mu + \lambda)$ algorithm. A more complete definition of the algorithm is presented in Fig 4.14. After the operators and evolution's numeric parameters are set in the population's counter loop, separate population's variation is performed. The algorithm's result is the best solution in the last generation.

The third approach generates a new offspring population having the size of λ by crossover, mutation, or reproduction (cloning) with probabilities P_{cx}, P_{mut} and $1 - (\mu + \lambda)$, respectively as shown in Fig. 4.14. This new generation is formed by selection *only* in the offspring population. Hereinafter, this algorithm is called (μ, λ) algorithm. A fuller algorithm description is presented by the flowchart in Fig. 4.15. It repeats the previous algorithm except for the way of selection that is performed from the newly created (changed) population and the preceding initial population.

We start the search for the GAs stop criterion with vehicle's path planning using the three approaches presented above. The task is solved by the planner based on the data from the technical vision system and is performed in real time and without mapping. However, in the experiments a simple map is used because it gives a better algorithm load for the individual's divergence in the evolution process. It should be mentioned that the map used is not a mapping procedure but is used for an initial tuning of algorithm in the beginning of the research. The map with circled obstacles is presented in Fig. 4.16 while Fig. 4.17 presents it in the form of optimized graph.

As we can see from Fig. 4.16, the modeling scene has four point obstacles located on the vehicle's way from the initial point $(x_0; y_0)$ to the goal point $(x_g; y_g)$ where $(x_0; y_0) = (0;0)$, $(x_g; y_g) = (k; m)$. The obstacles in Fig. 4.17 are presented by missing vertices (black rectangles). All the other vertices of the graph are connected by the edges. If it becomes necessary to account for the obstacles clearance radius, the vertices surrounding the obstacle are deleted from the graph.

The main parameters of the considered GA are presented in Table 4.1.

The results of basic algorithms functioning are presented in Figs. 4.18–4.24. Fig. 4.18 presents the path evolution process.

The graph weight change is presented in Fig. 4.19. The change of RMSE of the graph's weight is presented in Fig. 4.20. The solid line in Fig. 4.19 shows the change of the average weight value for all the population in the generation. The dashed line shows the value for the best individual in this generation.

As we see from the graphs presented in Figs. 4.19 and 4.20, starting from the 140th generation, the graph's weight and its RMSE change insignificantly. The steady state value of the graph's weight equals to 19 with oscillating average value. The RMSE of the population fitness reduces in evolution but during generations 120–160 it oscillates around the level of 2, being 10% of the steady state value of fitness.

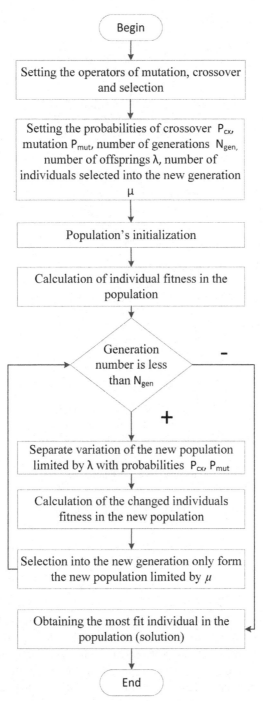

FIGURE 4.14 Flowchart of the genetic algorithm ($\mu + \lambda$).

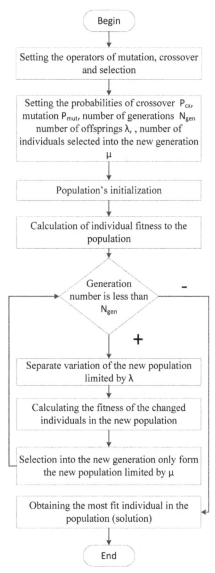

FIGURE 4.15 Flowchart of the genetic algorithm (μ, λ).

Thus, inside the boundary of the set number of generations, the process of genetic calculations doesn't have any visible evolution stop criteria. Noticing the declining character of the RMSE curve, the increase of the number of generations allows to state an obvious prevalence of maximally fit individuals, which, on the other hand, leads to significant growth of computational costs.

The evolution process of the considered graph from the point of view of the changed individuals is shown in Figs. 4.21—4.24. It should be mentioned that the tested algorithm type assumes consecutive application of crossover and

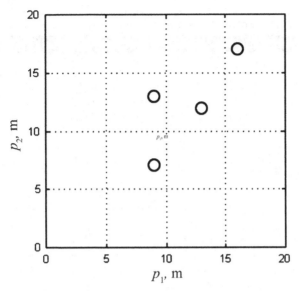

FIGURE 4.16 Region map.

mutation, and does not exclude the possibility of getting a new individual using both operators together. Each point at the graph 4.24 represents the number of fitness estimates changed in the new generation. The dotted line shows the average value of fitness estimates for the entire evolution interval. The line consisting vertical dashes represents a median. Since [33–35] the most computational load goes to the operations changing the individuals during software implementation, the solid line in Fig. 4.24 can be used for mediated estimation of the algorithm's resource intensity.

The results of $(\mu + \lambda)$ algorithm functioning are shown in Fig. 4.25.

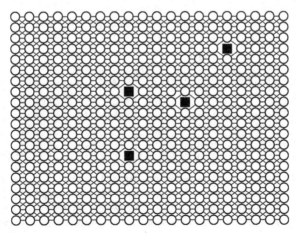

FIGURE 4.17 Region map in a graph form.

Table 4.1 Main Parameters of the Genetic Algorithms

Parameter	Basic Algorithm	$(\mu + \lambda)$ Algorithm	(μ,λ) Algorithm
Mutation type		Genes position change	
Mutation probability		0.2	
Crossover type		Two points	
Crossover probability		0.5	
Selection type		Tournament	
Population size, individual		100	
λ, individual		200	
μ, individual		100	
Generations number	300	200	150

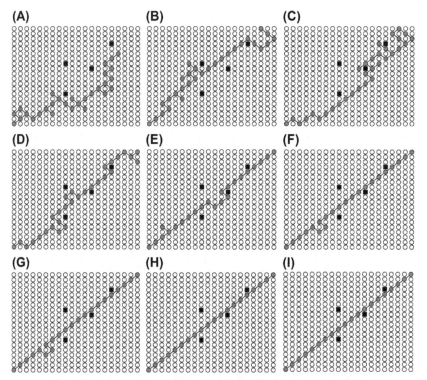

FIGURE 4.18 Vehicle's path evolution process: (A) generation 1, graph weight 32; (B) generation 40, graph weight 24; (C) generation 78, graph weight 24; (D) generation 120, graph weight 24; (E) generation 148, graph weight 23; (F) generation 173, graph weight 21; (G) generation 191, graph weight 21; (H) generation 206, graph weight 19; and (I) generation 216, graph weight 19.

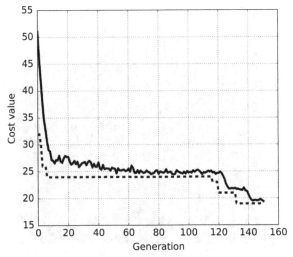

FIGURE 4.19 Evolution of average and minimal values of graph weight.

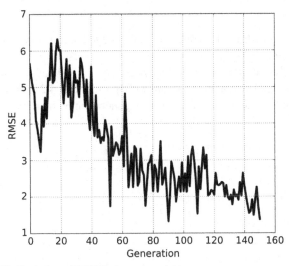

FIGURE 4.20 Evolution of RMSE of graph weight.

The process and quality of the GA ($\mu + \lambda$) functioning can be estimated looking at Figs. 4.26 and 4.27. In Fig. 4.26 the solid line represents the values of population fitness and the dashed line the change of the most fit individual. As we see from this figure, and from Fig. 4.27, the process converges faster than the traditional algorithm presented above. In the process of evolution the population degenerates two times—for the first time at the fitness level of 25 and second time at the level of 19. The second degeneration indicates the calculations convergence. The degeneration corresponds to the relatively small values

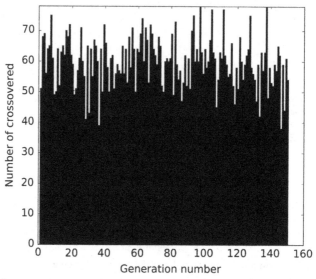

FIGURE 4.21 Number of crossed individuals. Number of individuals generated by crossover.

of RMSE reaching the values of 2 for the first degeneration and 1 for the second as demonstrated in Fig. 4.27.

The evolution process for the changed individuals is presented in Figs. 4.28–4.30. The columnar graphs in Figs. 4.28 and 4.29 show the number of crossed and mutated individuals in each generation with probabilities 0.5 and

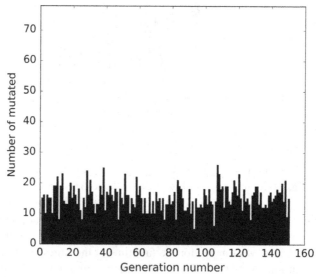

FIGURE 4.22 Number of mutated individuals. Number of individuals generated by mutation.

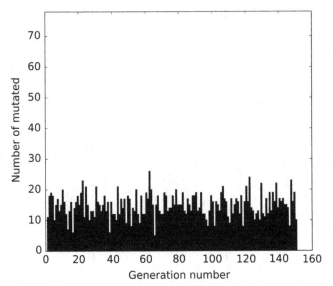

FIGURE 4.23 Number of crossed and mutated individuals.

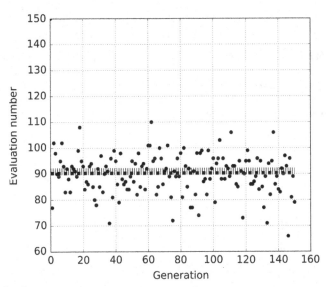

FIGURE 4.24 Number of fitness calculations of the new individuals.

0.2, respectively. The graph in Fig. 4.30 demonstrates the total number of new individuals for each generation. It should be mentioned that the number of new individuals for each generation is greater than the initial size of the generation itself. This ensures the necessary variability among the individuals ensuring the algorithm convergence.

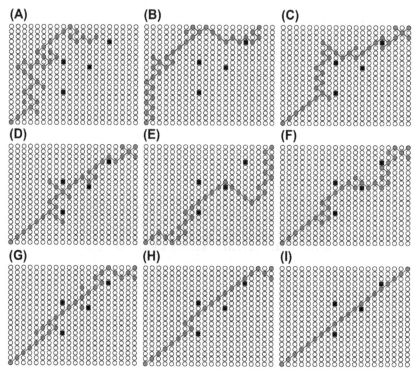

FIGURE 4.25 Path evolution with $(\mu + \lambda)$ algorithm: (A) generation 1, graph weight 30; (B) generation 20, graph weight 24; (C) generation 41, graph weight 24; (D) generation 61, graph weight 24; (E) generation 82, graph weight 24; (F) generation 103, graph weight 24; (G) generation 108, graph weight 24; (H) generation 111, graph weight 23; and (I) generation 143, graph weight 19.

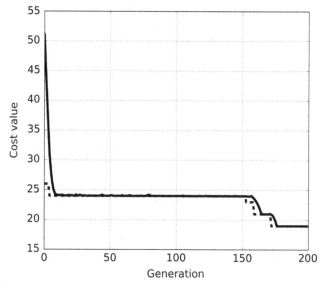

FIGURE 4.26 Average and minimal graph weight with $(\mu + \lambda)$ algorithm.

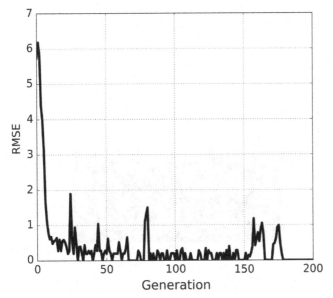

FIGURE 4.27 RMSE of the average and the minimal graph weight with a $(\mu + \lambda)$ algorithm.

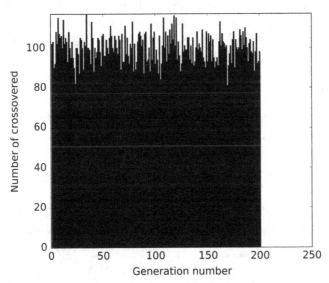

FIGURE 4.28 Number of crossed individuals with a $(\mu + \lambda)$ algorithm.

The results of (μ,λ) algorithm functioning for the path-finding problem are presented in Fig. 4.31.

Figs. 4.32–4.36 demonstrate the evolution process in the context of population change. In the performed experiment the process converges to the steady

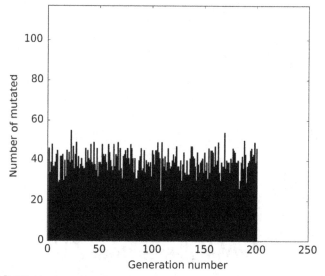

FIGURE 4.29 Number of mutated individuals with a $(\mu + \lambda)$ algorithm.

state value faster than the two previous cases. As we see in Fig. 4.32, the solid line representing the average value of population fitness matches the dashed line of the best individual fitness after passing the 63rd generation and practically remains constant. There is also degeneration at the two levels of fitness—24 and 19. As we see from Fig. 4.33, for the second degradation the value of RMSE is less than 1, which can be used as a stop criterion.

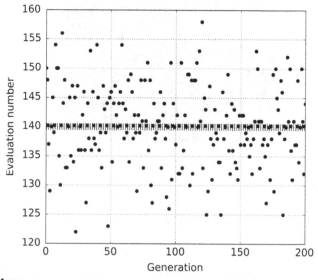

FIGURE 4.30 Number of fitness calculations with $(\mu + \lambda)$ algorithm.

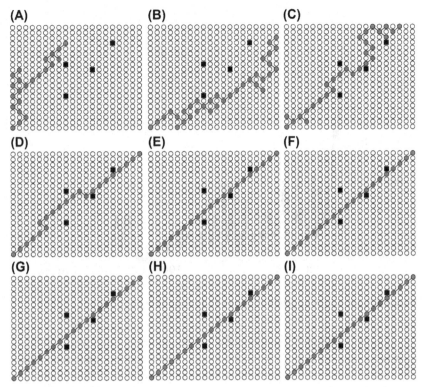

FIGURE 4.31 Path evolution with (μ,λ) algorithm: (A) generation 1, graph weight 36; (B) generation 22, graph weight 24; (C) generation 44, graph weight 24; (D) generation 66, graph weight 21; (E) generation 88, graph weight 19; (F) generation 110, graph weight 19; (G) generation 132, graph weight 19; (H) generation 154, graph weight 19; and (I) generation 176, graph weight 19.

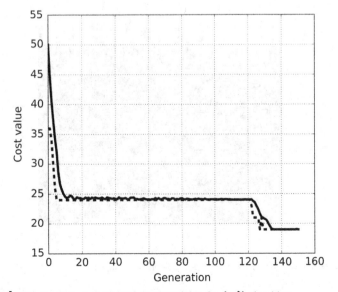

FIGURE 4.32 Average and minimal graph weights for (μ,λ) algorithm.

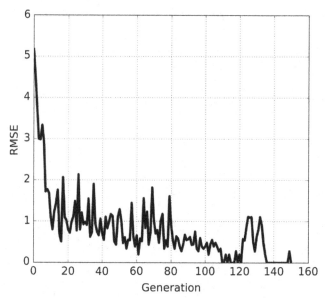

FIGURE 4.33 Graph weights RMSE with (μ, λ) algorithm.

As it follows from Figs. 4.34 and 4.35, the pattern of crossovers and mutations is the same as in the previous case since the population variation algorithm is the same, and so are the probability values for crossovers and mutations. The overall picture of population variation is presented in Fig. 4.36. The computational load level used for fitness estimation is approximately the same as for the previous experiment.

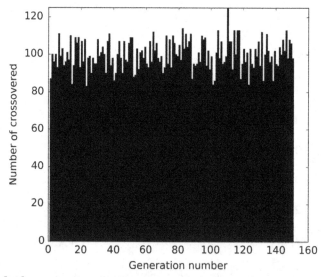

FIGURE 4.34 Number of crossed individuals in (μ, λ) algorithm.

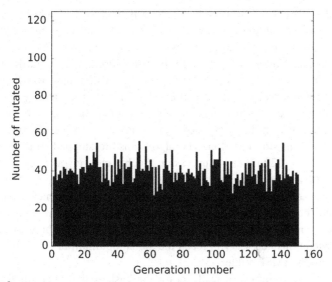

FIGURE 4.35 Number of mutated individuals in (μ,λ) algorithm.

From the performed experiments a conclusion can be made that $(\mu + \lambda)$ algorithm is the most efficient and has the most obvious convergence criteria to the steady state solution in the form of RMSE of the population fitness oscillating at a level below 1. However, it should be mentioned that the used stop criterion is sufficient for the preliminary estimations of the algorithm's functionality but can't give a guaranteed result upon being implemented as a part of an autonomous control system.

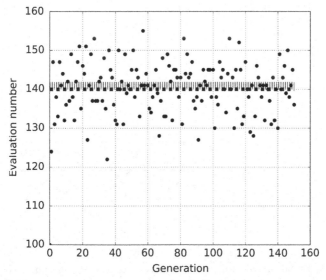

FIGURE 4.36 Number of fitness calculations in (μ,λ) algorithm.

On the other hand in the research described above it was mentioned that convolution of evolutionary calculations takes essentially different number of generations for the same algorithm with same parameters. This is explained by a significant random component of the calculations. Hence, for determining the real algorithm's efficiency and for calculation of fitness RMSE numeric value serving as a calculation stop criterion, the most representative result samples were analyzed. For this purpose, a series of experiments were performed with number of tests for each algorithm equal to 100. The algorithms were modified only in stop criterion. The algorithms are stopped if a steady state value is achieved 20 times in succession. Such a choice was made looking at the algorithm's functioning results presented in Figs. 4.19, 4.26 and 4.32. Particularly, the calculations should not diverge after convergence to a constant value and the fitness RMSE goes down indicating a global optimum. The mentioned tuning criterion will allow us to analyze the algorithms for the number of generations necessary to reach the global optimum of generations number and fitness RMSE values.

The results of statistical analysis for the basic genetic path-finding algorithm are presented in Figs. 4.37–4.42.

The dotted markers in Fig. 4.37 illustrate the required number of generations for each launch of algorithm. The solid line corresponds to the arithmetic mean value of the required number of generations. The histogram in Fig. 4.38 shows the distribution of the number of experiments in relation to the required number of generations having a peak of the mean value at the level of 150 generations. An asymmetric bias to the right is explained by single instances of the

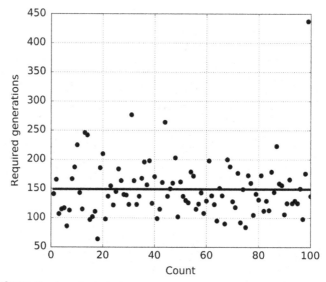

FIGURE 4.37 Number of calculations necessary for calculations convergence of the basic algorithm.

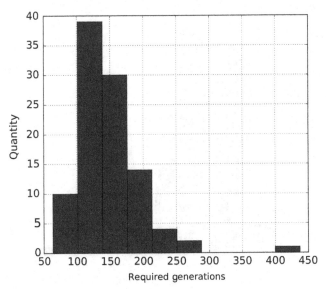

FIGURE 4.38 Histogram of the number of generations necessary for calculations convergence of the basic algorithm.

most prolonged launches, which is a drawback in the considered algorithm type.

Figs. 4.39 and 4.40 demonstrate the state of the data population fitness RMSE used for the stop criterion identification in the last implementation of the algorithm. The dotted markers in Fig. 4.39 illustrate the maximal values

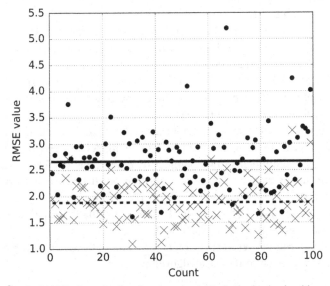

FIGURE 4.39 RMSE of maximal and average weights in the basic algorithm.

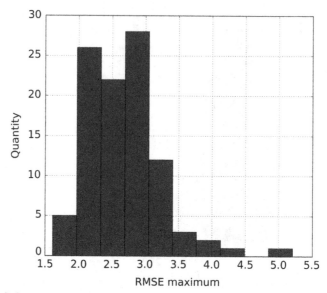

FIGURE 4.40 Histogram of maximal population fitness RMSE for the basic algorithm.

of the fitness RMSE for the last 20 successive generations where each generation ensures an optimal solution. The dotted line is at the level of the mean value of this parameter for all the experiment series. The histogram in Fig. 4.40 reveals a peak of the number of experiments at the value of maximal RMSE equal to approximately 2.5. The asymmetry of the distribution to the large number of RMSE indicated the possibility of calculations convergence with a large divergence in the population. And since the large divergence is typical for the populations without an optimal individual, it becomes difficult to determine the calculations stop criterion.

The time indices for the basic algorithm are presented in Figs. 4.41 and 4.42. The dotted markers in Fig. 4.41 illustrate results for each of the experiments while the solid line shows the average value for all the series of 100 experiments. Since the algorithm functioning duration is proportional to the necessary number of generations, the functioning histogram demonstrates an asymmetry analogous to the previous histograms.

The results of statistical research for the series of $(\mu + \lambda)$ algorithms for the path-finding task are presented in Figs. 4.43–4.48. As we see from Figs. 4.43 and 4.44, the requirements to the number of generations for the $(\mu + \lambda)$ algorithms are higher than for the basic algorithm but the distribution pattern is more symmetric and has a more clearly marked peak.

As shown in Figs. 4.45 and 4.46, RMSE of population fitness are located more densely around the mean value with singular surges.

The time criteria presented in Figs. 4.47 and 4.48 indicate worse performance in comparison to the basic algorithm.

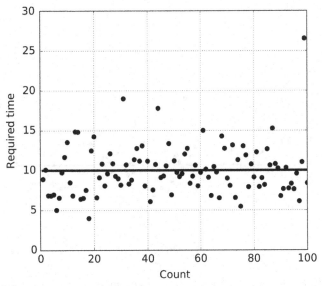

FIGURE 4.41 Basic algorithm functioning time.

The results of (μ,λ) algorithm experiment series are presented in Figs. 4.49–4.54. The required number of generations distribution has an asymmetrical form but the average value is higher than in both preceding algorithms.

The maximal fitness RMSE has a higher dispersion than one for $(\mu + \lambda)$ algorithm. However, there is deformation of the RMSE distribution to the right

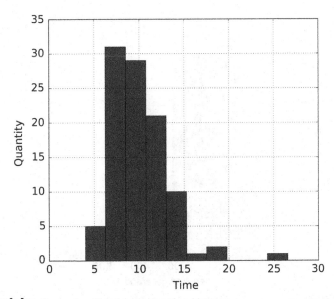

FIGURE 4.42 Basic algorithm functioning time histogram.

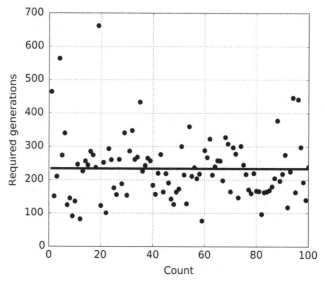

FIGURE 4.43 Generations number of $(\mu + \lambda)$ algorithm.

side. The graphs presented in Figs. 4.53 and 4.54 indicate better functioning time than that of $(\mu + \lambda)$ algorithm but worse than the time of the basic algorithm.

For the purpose of algorithm selection, let us address the resulting functioning indices collected for the series of experiments in Table 4.2.

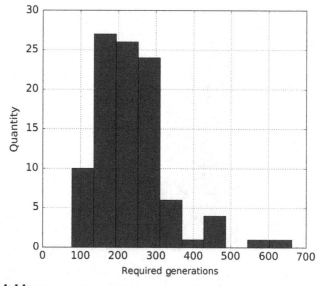

FIGURE 4.44 Number of generations histogram $(\mu + \lambda)$ algorithm.

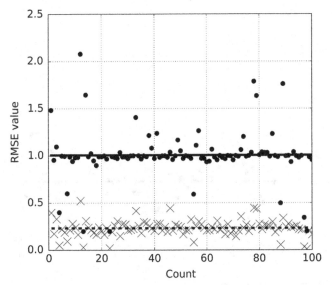

FIGURE 4.45 RMSE of maximal and mean values of $(\mu + \lambda)$ algorithm.

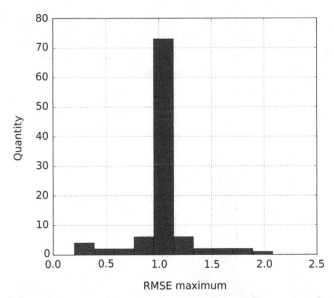

FIGURE 4.46 Histogram of maximal RMSE of population fitness for $(\mu + \lambda)$ algorithm.

As we see from Table 4.2, the best speed is reached by the basic algorithm. But it loses in convergence speed to $(\mu + \lambda)$ and (μ,λ) algorithms. The algorithm $(\mu + \lambda)$ is the slowest one but is the fastest in convergence. Therefore, for the planning system (μ,λ) algorithm was selected. Essential functioning

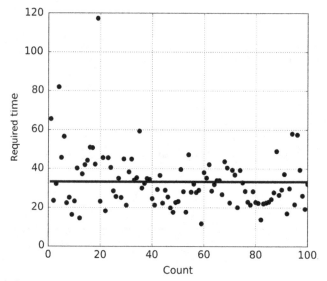

FIGURE 4.47 ($\mu + \lambda$) Algorithm functioning time.

time reduction is achieved by code optimization by disengaging parallel computations in the implemented algorithm and by selection of the population size and probabilities of crossover and mutations. The final functioning time is around 1 s, which is confirmed by the modeling results.

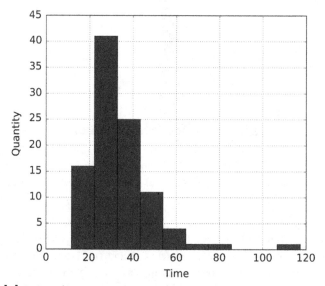

FIGURE 4.48 ($\mu + \lambda$) Algorithm functioning time histogram.

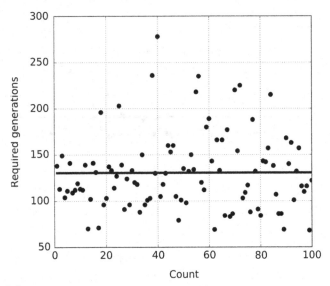

FIGURE 4.49 (μ,λ) Algorithm generations number.

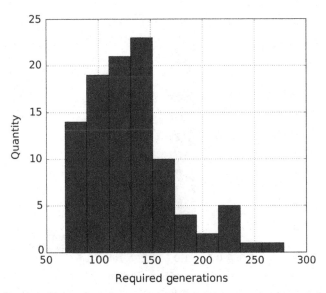

FIGURE 4.50 (μ,λ) Algorithm generations number histogram.

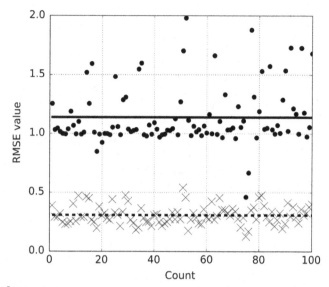

FIGURE 4.51 RMSE of maximal and mean values of (μ, λ) algorithm.

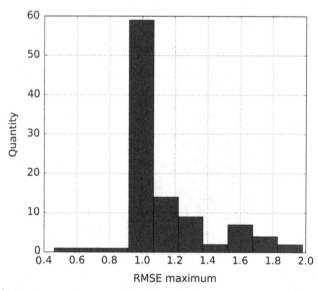

FIGURE 4.52 Histogram of maximal RMSE of population fitness for (μ, λ) algorithm.

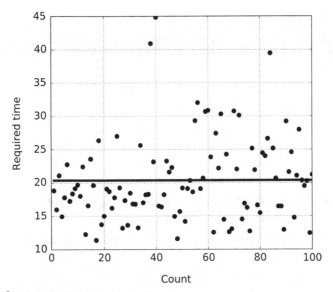

FIGURE 4.53 (μ, λ) Algorithm functioning time.

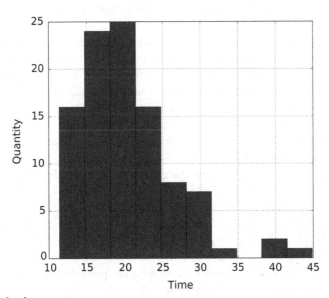

FIGURE 4.54 (μ, λ) Algorithm functioning time histogram.

Table 4.2 Algorithms Functioning Indicators for Each Series of 100 Experiments

Indicator, Mean Value	Basic Algorithm	$(\mu + \lambda)$ Algorithm	(μ, λ) Algorithm
Generations	148.00	211.72	132.27
Time	10.56	28.32	20.78
Average fitness RMSE	2.61	1.05	1.24
Maximal fitness RMSE	5.32	2.1	2.0

4.4 MODELING RESULTS OF USING GENETIC ALGORITHMS FOR PATH FINDING

4.4.1 Imitation Modeling Results Without Mapping

This section presents the imitation modeling results for the vehicle described in Chapter 1 using the algorithm described above without mapping. That is, the planner receives the information about the obstacles located in the sensor's visibility range at the current modeling step. In the next step, the planner obtains new data discarding the information received at the previous step. The planning algorithm uses a graph with a maximal number of nodes along abscissa equal to $k = 20$ and a maximal number of nodes along the ordinate equal to $m = 20$. The robot moves from the initial point (0; 0) to the goal point (20; 20). The sampling interval is 1 m for each coordinate axis.

The graphs show the robot's motion path and circular obstacles. The cross markers indicate the points of the planned trajectory generated by the planner.

The vehicle's motion modeling results for the first scene are presented in Fig. 4.55. The robot successfully avoids all the obstacles and stops at the goal point. The succession of path points is presented in Table 4.3.

The control system's motion performance criteria: safety criterion Y, $S_m = 0.262$; path length, $P_L = 29.5436$; task completion time $t_m = 36.36$; and mission success coefficient $F = 1$.

The modeling results for scene 2 are presented in Fig. 4.56. The succession of path points is presented in Table 4.4. As we see from Fig. 4.56 and Table 4.4, the robot successfully avoids all the obstacles and stops at the goal point.

The performance criteria for motion modeling in scene 2: safety criterion $S_m = 0.10151$; path length $P_L = 29.5869$; task completion time $t_m = 36.18$; and mission success coefficient $F = 1$.

The motion modeling results for scene 3 are presented in Fig. 4.57. The planned trajectory is presented in Table 4.5.

At time $t_i = 31.86$, the robot is located at the intermediate planned point (15.5; 14.5) with an orientation angle equal to approximately 45 degrees. Then it proceeds to the next point (14.5; 14.5) planned according to the scanner data that detected no obstacle in its visibility range. This leads to collision of robot and obstacle, and planner inoperability explained by the fact that robot

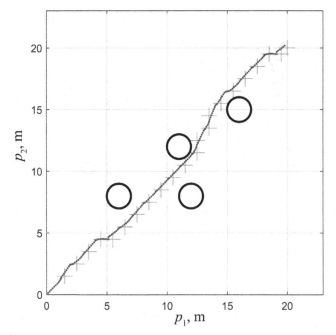

FIGURE 4.55 Scene 1 modeling results.

gets into a position corresponding to an obstacle graph vertex—a vertex that is not connected to any other vertex by edges. The total path length P_L until the collision equals to 22.7577 m.

The modeling results for the vehicle moving in scene 4 are presented in Fig. 4.58. Similar to the case of scene 3, the robot collides with an obstacle and becomes unable to continue motion for the same reason as in the previous case. At the collision moment $t_i = 29.52$ the path length P_L was equal to 21.736. The motion path is presented in Table 4.6.

The motion modeling results for the scene 5 are presented in Fig. 4.59. In this case a collision took place as well. At the collision moment $t_i = 28.44$ the path length P_L was equal to 20.5080 m. The succession of points passed by the path planner to the control system is presented in Table 4.7.

Table 4.3 Motion Path for Scene 1

										Point Num.												
Coord.	1	2	3	4	5	6	7	8	9	10	11	12	13	14	15	16	17	18	19	20	21	22
x, m	1.5	2.5	3.5	4.5	5.5	6.5	7.5	8.5	9.5	10.5	11.5	12.5	12.5	13.5	13.5	14.5	15.5	16.5	17.5	18.5	19.5	20.5
y, m	1.5	2.5	3.5	4.5	4.5	5.5	6.5	7.5	8.5	9.5	10.5	11.5	12.5	13.5	14.5	15.5	16.5	17.5	18.5	19.5	19.5	20.5

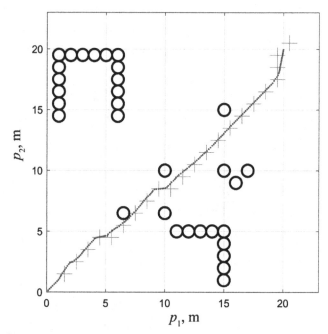

FIGURE 4.56 Scene 2 modeling results.

Based on these modeling results we can make a conclusion that the proposed method used without mapping gives unsatisfactory results when working with complicated obstacles of L- or Π-type.

The mentioned drawback, which comes out in path planning for the environment with complicated obstacles, is generally inherent to the common qualities of GAs. In order to avoid this drawback, a mapping procedure is incorporated into the planning process. It is performed in real time during the robot's motion.

The resulting performance criteria of the path planning system for all the five scenes are combined in Table 4.8.

Table 4.4 Motion Path for Scene 2

Coord.	Point Num.																					
	1	2	3	4	5	6	7	8	9	10	11	12	13	14	15	16	17	18	19	20	21	22
x, m	1.5	2.5	3.5	4.5	5.5	6.5	7.5	8.5	9.5	10.5	11.5	12.5	13.5	14.5	15.5	16.5	17.5	18.5	19.5	19.5	19.5	20.0
y, m	1.5	2.5	3.5	4.5	4.5	5.5	6.5	7.5	8.5	8.5	9.5	10.5	11.5	12.5	13.5	14.5	15.5	16.5	17.5	18.5	19.5	20.0

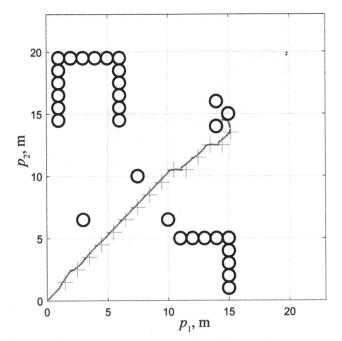

FIGURE 4.57 Scene 3 modeling results.

4.5 IMITATION MODELING RESULTS FOR PATH PLANNING WITH MAPPING

The path planner's failures described in the previous section can be avoided by introduction of mapping. In other words, any obstacle that gets into the scanner's visibility range is stored in memory and is used for future path planning.

A robot with a mapping genetic path planner was modeled in five different scenes. The modeling results for the scene 1 are presented in Fig. 4.60.

As seen from Fig. 4.60, the robot successfully completes its mission with the following performance criteria: safety criterion $S_m = 0.3294$; path length $P_L = 29.4225$; task completion time $t_m = 34.20$; and mission success coefficient $F = 1$. The scene 1 path is presented in Table 4.9.

The imitation modeling results for the robot moving in the scene 2 are presented in Fig. 4.61.

Table 4.5 Motion Path for Scene 3

	Point Num.															
Coord.	1	2	3	4	5	6	7	8	9	10	11	12	13	14	15	16
x, m	1.5	2.5	3.5	4.5	5.5	6.5	7.5	8.5	9.5	10.5	11.5	12.5	13.5	14.5	15.5	14.5
y, m	1.5	2.5	3.5	4.5	5.5	6.5	7.5	8.5	9.5	10.5	10.5	11.5	12.5	12.5	13.5	14.5

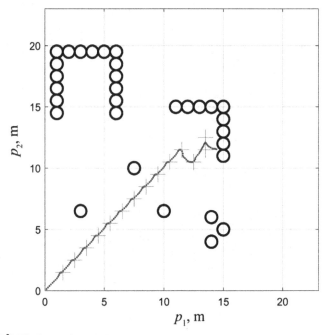

FIGURE 4.58 Scene 4 modeling results.

Table 4.6 Motion Path for Scene 4

							Point Num.								
Coord.	1	2	3	4	5	6	7	8	9	10	11	12	13	14	15
x, m	1.5	2.5	3.5	4.5	5.5	6.5	7.5	8.5	9.5	10.5	11.5	12.5	13.5	13.5	14.5
y, m	1.5	2.5	3.5	4.5	5.5	6.5	7.5	8.5	9.5	10.5	11.5	10.5	11.5	12.5	11.5

The performance criteria for the scene 2: $S_m = 0.3446$; path length $P_L = 29.4713$; task completion time $t_m = 34.38$; and mission success coefficient $F = 1$. The motion path is presented in Table 4.10.

The imitation modeling results for the scene 3 are presented in Fig. 4.62.

The performance criteria for the scene 3: $S_m = 1.0558$; path length $P_L = 29.3445$; task completion time $t_m = 34.38$; and mission success coefficient $F = 1$.

The motion path is presented in Table 4.11.

The imitation modeling results for the scene 4 are presented in Fig. 4.63.

The performance criteria for the scene 4: path length $P_L = 34.1712$; task completion time $t_m = 55.62$; and mission success coefficient $F = 1$. The planned path is presented in Table 4.12.

The imitation modeling results for the scene five are presented in Fig. 4.64.

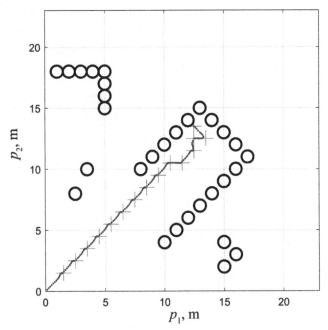

FIGURE 4.59 Scene 5 modeling results.

Table 4.7 Motion Path for Scene 5

Coord.	Point Num.														
	1	2	3	4	5	6	7	8	9	10	11	12	13	14	15
x, m	1.5	2.5	3.5	4.5	5.5	6.5	7.5	8.5	9.5	10.5	11.5	12.5	12.5	13.5	12.5
y, m	1.5	2.5	3.5	4.5	5.5	6.5	7.5	8.5	9.5	10.5	10.5	11.5	12.5	12.5	13.5

Table 4.8 Performance Criteria for Path Planning Without Mapping

Parameter	Scene #1	Scene #2	Scene #3	Scene #4	Scene #5
S_m	0.2662	0.1015	0	0	0
P_L	29.5436	29.5869	–	–	–
t_m	36.36	36.18	–	–	–
F	1	1	1	1	1

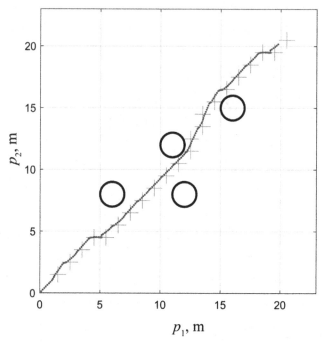

FIGURE 4.60 Scene 1 modeling results.

Table 4.9 Motion Path for Scene 1 With Mapping

Coord.	Point Num.																					
	1	2	3	4	5	6	7	8	9	10	11	12	13	14	15	16	17	18	19	20	21	22
x, m	1.5	2.5	3.5	4.5	5.5	5.5	6.5	7.5	8.5	9.5	11.5	12.5	12.5	13.5	13.5	14.5	15.5	16.5	17.5	18.5	19.5	20.5
y, m	1.5	2.5	3.5	4.5	5.5	6.5	7.5	9.5	10.5	11.5	9.5	11.5	12.5	13.5	14.5	15.5	16.5	17.5	18.5	19.5	19.5	20.5

The performance criteria for the scene 5: $S_m = 1.0557$; path length $P_L = 42.6787$; task completion time $t_m = 75.42$; and mission success coefficient $F = 1$. The motion path is presented in Table 4.13.

The performance criteria calculated using the experiment results for all the five scenes with mapping are presented in Table 4.14.

Thus, for the presented GAs to become capable of successful path planning, it is necessary to use mapping during the vehicle's motion using the data coming from technical vision system. This does not increase the task completion time.

The path planning method presented in this chapter is compared to other methods in the summary to this monograph.

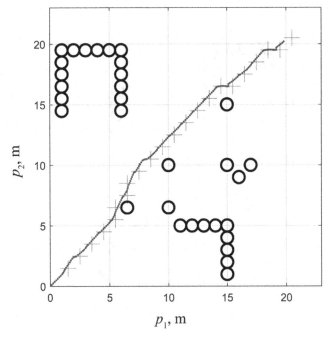

FIGURE 4.61 Modeling results for scene 2.

Table 4.10 Motion Path for Scene 2 With Mapping

Coord.	Point Num.																					
	1	2	3	4	5	6	7	8	9	10	11	12	13	14	15	16	17	18	19	20	21	22
x, m	1.5	2.5	3.5	4.5	5.5	6.5	7.5	8.5	9.5	10.5	11.5	10.5	11.5	12.5	13.5	14.5	15.5	16.5	17.5	18.5	19.5	20.5
y, m	1.5	2.5	3.5	4.5	4.5	5.5	6.5	7.5	8.5	8.5	9.5	12.5	13.5	14.5	15.5	16.5	16.5	17.5	18.5	19.5	19.5	20.5

4.6 SUMMARY

In this chapter GAs were applied for vehicle path planning in the environment with obstacles using the example of a wheeled cart.

An algorithm implementation type was selected having the best resource intensity and convergence speed. An evolution stop criterion for a particular implementation was found. The criterion selection and its numeric values were validated statistically.

The modeling results have shown the necessity for mapping if the vehicle is moving around complicated obstacles. Genetic planning based on GAs without mapping leads to a local minimum.

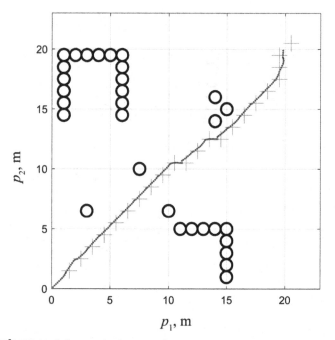

FIGURE 4.62 Modeling results for scene 3.

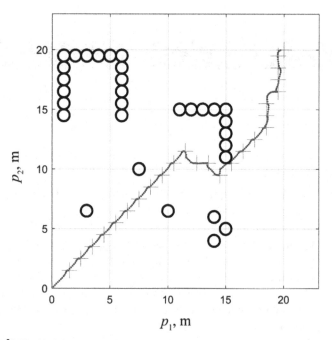

FIGURE 4.63 Modeling results for scene 4.

Table 4.11 Motion Path for Scene 3 With Mapping

											Point Num.											
Coord.	1	2	3	4	5	6	7	8	9	10	11	12	13	14	15	16	17	18	19	20	21	22
x, m	1.5	2.5	3.5	4.5	5.5	6.5	7.5	8.5	9.5	10.5	11.5	12.5	13.5	14.5	15.5	16.5	17.5	18.5	19.5	19.5	19.5	20.5
y, m	1.5	2.5	3.5	4.5	5.5	6.5	7.5	8.5	9.5	10.5	10.5	11.5	12.5	12.5	13.5	14.5	15.5	16.5	17.5	18.5	19.5	20.5

Table 4.12 Motion Path for Scene 4 With Mapping

												Point Num.												
Coord.	1	2	3	4	5	6	7	8	9	10	11	12	13	14	15	16	17	18	19	20	21	22	23	24	25
x, m	1.5	2.5	3.5	4.5	5.5	6.5	7.5	8.5	9.5	10.5	11.5	12.5	13.5	14.5	15.5	16.5	17.5	18.5	18.5	18.5	19.5	19.5	19.5	19.5	20.5
y, m	1.5	2.5	3.5	4.5	5.5	6.5	7.5	8.5	9.5	10.5	11.5	10.5	10.5	9.5	10.5	11.5	12.5	13.5	14.5	15.5	16.5	17.5	18.5	19.5	20.5

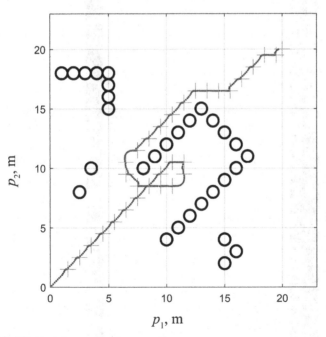

FIGURE 4.64 Modeling results for scene 5.

Table 4.13 Motion Path for Scene 5 With Mapping

Coord.	Point Num.																															
	1	2	3	4	5	6	7	8	9	10	11	12	13	14	15	16	17	18	19	20	21	22	23	24	25	26	27	28	29	30	31	32
x, m	1.5	2.5	3.5	4.5	5.5	6.5	7.5	8.5	9.5	10.5	11.5	11.5	10.5	9.5	8.5	7.5	6.5	6.5	7.5	8.5	9.5	10.5	11.5	12.5	13.5	14.5	15.5	16.5	17.5	18.5	19.5	20.0
y, m	1.5	2.5	3.5	4.5	5.5	6.5	7.5	8.5	9.5	10.5	10.5	9.5	8.5	8.5	8.5	8.5	9.5	10.5	11.5	12.5	13.5	14.5	15.5	16.5	16.5	16.5	16.5	17.5	18.5	19.5	19.5	20.0

Table 4.14 Performance Criteria for the Experiments With Mapping

Parameter	Scene 1	Scene 2	Scene 3	Scene 4	Scene 5
S_m	0.3294	0.3446	1.0558	0.0393	1.0557
P_L	29.4225	29.4713	29.3445	34.1712	42.6787
t_m	34.20	34.38	34.38	55.62	75.42
F	1	1	1	1	1

REFERENCES

[1] M. Barbehenn, S. Hutchinson, Efficient search and hierarchical motion planning by dynamically maintaining single-source shortest paths trees, IEEE Transactions on Robotics and Automation 11 (2) (1995) 198–214.

[2] J.R. Evans, E. Minieka, Optimization Algorithms for Networks and Graphs, second ed., M. Dekker, New York, 1992.

[3] C.H. Papadimitriou, P. Raghavan, M. Sudan, H. Tamaki, Motion planning on a graph, FOCS 94 (1994) 511–520.

[4] B. Dancila, R. Botez, D. Labour, Altitude optimization algorithm for Cruise, constant speed and level flight segments, in: AIAA Guidance, Navigation, and Control Conf., Amer. Inst. of Aeronautics and Astronautics, 2012.

[5] R. Dancila, R.M. Botez, S. Ford, Fuel burn and emissions evaluation for a missed approach procedure performed by a B737-400, in: Aviation Technology, Integration, and Operations Conf., Amer. Inst. of Aeronautics and Astronautics, 2013.

[6] R.S. Felix-Patron, R. Botez, D. Labour, New altitude optimisation algorithm for the flight management system CMA-9000 improvement on the A310 and L-1011 aircraft, Royal Aeronautical Society 117 (2013) 787–805.

[7] S. Liden, Optimum Cruise Profiles in the Presence of Winds, Proc. of Digital Avionics Systems Conf., IEEE/AIAA 11th, Seattle, WA, 1992, pp. 254–261.

[8] S. Liden, Practical Considerations in Optimal Flight Management Computations, American Control Conf., Boston, MA, USA, 1985, pp. 675–681.

[9] F.L.L. bo Medeiros, J.D.S. es da Silva, A. Rocha Costa, et al., A Dijkstra Algorithm for Fixed-wing UAV Motion Planning Based on Terrain Elevation, Advances in Artificial Intelligence SBIA 2010, vol. 6404, Springer, Berlin–Heidelberg, 2010, pp. 213–222.

[10] E. Rippel, A. Bar-Gill, N. Shimkin, Fast graph-search algorithms for general-aviation flight trajectory generation, Journal of Guidance, Control, and Dynamics 28 (2005) 801–811.

[11] S. Sidibe, R. Botez, Trajectory optimization of FMS-CMA 9000 by dynamic programming, in: ASI AÉ RO 2013 Conf., 60th Aeronautics Conf. and AGM Toronto, Canada, 2013.

[12] R. Felix-Patron, A. Kessaci, R.M. Botez, Flight trajectories optimization under the influence of winds using genetic algorithms, in: AIAA Guidance, Navigation, and Control (GNC) Conf., Amer. Inst. of Aeronautics and Astronautics, Boston, USA, 2013.

[13] S. Asadi, V. Azimirad, A. Eslami, A. Ghanbari, A novel global optimal path planning and trajectory method based on adaptive dijkstra-immune approach for mobile robot, Advanced Intelligent Mechatronics (AIM), in: IEEE/ASME Intern. Conf, 2011, pp. 1093–1098.

[14] K. Berzh, Graph Theory and its Applications, Foreign literature, Moscow, 1962.

[15] I. Grossman, V. Magnus, Groups and Their Graphs, Mir Publishers, Moscow, 1971.

[16] V.A. Evstigneev, Graphs Theory Application for Programming, Nauka, Moscow, 1985.

[17] V.A. Emelichev, O.I. Melnikov, Lectures on Graph Theory, Nauka, Moscow, 1990.

[18] A.A. Zykov, Graph Theory Basis, Nauka, Moscow, 1987.

[19] N. Kristofides, Graph Theory, Algorithmic Approach, Mir Publishers, Moscow, 1978.

[20] F.A. Novikov, Discrete Mathematics for Programmers: College Textbook, third ed., Piter, Saint Petersburg, 2009.

[21] O. Ore, Graph Theory, Nauka, Moscow, 1980.

[22] R. Willson, Introduction to the Graphs Theory, Mir Publishers, Moscow, 1977.

[23] F. Harari, Graphs Theory, Mir Publishers, Moscow, 1973.

[24] J. Bang-Jensen, G. Gutin, Digraphs: Theory, Algorithms and Applications, Springer Monographs in Mathematics, Springer-Verlag, 2000.

[25] J.A. McHugh, Algorithmic Graph Theory, first ed., Prentice-Hall, Inc., Upper Saddle River, NJ, USA, 1990.

[26] L.A. Gladkov, V.V. Kurejchik, V.M. Kurejchik, Bioinspired Methods in Optimization, Fizmatlit, Moscow, 2009.

[27] L.A. Gladkov, V.V. Kurejchik, V.M. Kurejchik, Genetic Algorithms: Tutorial, second ed., Fizmatlit, Moscow, 2006.

[28] V.V. Emellyanov, V.V. Kurejchik, V.M. Kurejchik, Theory and Practice of Evolutional Modeling, Fizmatlit, Moscow, 2003.

[29] V.M. Kurejchik, B.K. Lebedev, O.K. Lebedev, Search Adaptation: Theory and Practice, Fizmatlit, Moscow, 2006.

[30] D. Rutkovskaya, M. Pilihjskij, L. Rutkovskij, Neural Networks, Genetic Algorithms and Fuzzy Systems, second ed., Hot line—Telecom, Moscow, 2008.

[31] Y.A. Skobczov, Evolutionary Computation Basis, Donetsk National Technical University, Donetsk, 2008.

[32] T. Bäck, D.B. Fogel, Z. Michalewicz, Evolutionary Computation 1: Basic Algorithms and Operators, 2000.

[33] F.M. De Rainville, F.A. Fortin, M.A. Gardner, M. Parizeau, C. Gagné, DEAP: enabling nimber evolutionss, SIGEvolution 6 (2) (2014) 17—25.

[34] F.M. De-Rainville, F.A. Fortin, M.A. Gardner, DEAP: a python framework for evolutionary algorithms, companion, in: Proc. of the Genetic and Evolutionary Computation Conf., 2012.

[35] F.A. Fortin, F.M. De Rainville, M.A. Gardner, DEAP: evolutionary algorithms made easy, Journal of Machine Learning Research 13 (2012) 2171—2175.

Chapter | Five

Graphic-Analytical Approaches to Vehicle's Motion Planning

V. Soloviev, I. Shapovalov, V. Finaev
Southern Federal University, Taganrog, Russia

5.1 POTENTIAL-FIELD METHOD IN VEHICLES CONTROL

The potential fields method has different names: "artificial potential fields", "virtual force field — VFF", "vector field histogram — VFH" etc. The method's idea is that paths are planned as a solution result of a special "motion equation" that includes "goal's attractive force", "obstacle's repulsive force" and other forces.

5.1.1 Fundamentals

According to Ref. [1] the idea of motion in a field of "informational" (virtual) forces is related to the work of Kurt Lewin—a pioneer of Gestalt psychology. In 1930—40 he proposed to use the concept of a physical field for description of behavior and conflict situations taking place during interaction of a person with the environment. The modern psychologists criticize K. Lewin for physicalism of his concept, making accents on the dynamic aspect at the expense of content and other things. However, K. Lewin's works reveal a lot of interesting things. The experimental data confirming the developed concept was obtained from observing children and was documented by filming. It is possible to draw a parallel between the Lewin's concept of physical field and potential field used for vehicle control.

According to Ref. [1] at the initial stage of the research performed at the Institute of Applied Mathematics of the Russian Academy of Sciences, the obstacles had the form of circles. The results obtained in Refs. [2,3] approved such representation of obstacles as the main. The attractive force was considered to be constant and directed toward the goal point. The repelling force f_i of ith obstacle depended on the argument R_i/r_i, where R_i is the circumference radius around the ith obstacle, r_i, distance from the ith obstacle circle center to the

185

Path Planning for Vehicles Operating in Uncertain 2D Environments. http://dx.doi.org/10.1016/B978-0-12-812305-8.00005-3

vehicle. The force f_i was directed away from the circle center. The trajectory (locomotion) was obtained as a result of integration of the second order equations.

During the research it was also found that the inertia inherent in the used model leads to a dangerous trajectory.

In order to avoid this drawback and make the method useful for the cases when other methods of obstacle contours approximation are used, the following actions were undertaken. First, the equations of the first order were used (i.e., the acting forces determine speed, analogous to the gradient descent method). Second, the value of the repelling force depends only on the distance to the obstacle, and its direction was determined at the closest point on the obstacle's surface. The obstacle is considered to have a shape with an arbitrary contour.

The research done by Refs. [2,3] for several obstacles showed that the best results are obtained if the repulsive functions of type x^{-k} and e^{-cx} are used, since they decrease fast with the growing distance to the obstacle. Changing the parameters k and c, when calculating the repulsive forces, allowed the authors to solve the tasks of obstacle avoidance and coordinated motion of several vehicles. Introduction of a delay made it possible to organize a "follow-me" mode.

If the environment with many obstacles is structured, e.g., the obstacles are joined into groups with noncrossing convex contours, the repulsive force can be calculated for all such groups of obstacles.

Hence, in the potential-field method, generalized models of environment are used enabling us to solve the planning task using minimal amount of information.

5.1.2 Survey of the Known Applications of the Potential-Fields Method

The potential-fields method dates back to 1950 as we can see in Ref. [4]. The results of detailed research performed in the Institute of Applied Mathematics of the Russian Academy of Sciences in USSR in 1974 are published in Ref. [5]. A pioneering work on mobile robot control using the ideas of "force field" method was published in 1984 [6]. In Ref. [7] there are references to other works published in 1985—86.

Analyzing other works for potential-fields method, two interesting directions can be distinguished.

The first one defines a potential field in a way to avoid stable equilibrium points totally. Obviously this task has a general solution. For example, a potential function at the point x equal to the length of the allowed path to the goal sets such a field. However, it is hard to calculate this function in a general case.

Koditschek and his co-authors proposed an interesting approach as a solution of this task [2,3]. In the beginning the obstacles are represented by circles and the resultant force is determined not as a sum of repulsive forces but as their product. This eliminates the force field equilibrium points. After that a transition is made to convex polygons and more complicated geometric representations. The generality of the obtained force fields is preserved.

However, it should be mentioned that such field representation for more general cases leads to paths that are not smooth and the local minima are also possible. Most likely for this reason the authors of Refs. [2,3] do not demonstrate the built motion paths.

The second direction is representing the vehicle not by a point but by a line segment or a rectangle [8]. This allows us to calculate both the resultant force and the moments of forces controlling the vehicle's orientation.

In Ref. [8] it is proposed to sum the attracting and repelling fields. The vehicle in this case is a multilink manipulator. The position, where the vehicle is to be moved to, is the center of the control system's attracting field and the obstacles are the center of repulsive field. Therefore, the vehicle is moving in the field of potential forces. In order to avoid the obstacles, a resulting repulsive field is calculated for each obstacle and then the total field of all obstacles is calculated.

The obstacles are represented by the equations of the following form

$$\left(\frac{x}{a}\right)^{2n} + \left(\frac{y}{b}\right)^{2n} + \left(\frac{z}{c}\right)^{2n} = 1 \tag{5.1}$$

where x, y, z are the coordinates of an arbitrary point on the obstacle; a, b, c constants determining the shape of the obstacle's surface; and n, positive number. The obstacle's closest point is selected to determine the vehicle's motion direction.

Ref. [8] gives the analytical equations for calculating the shortest distance between any of the manipulator links and the obstacles of a preset type.

The value of the field of attraction of the vehicle to the goal is determined by the following equation

$$U_{X_d} = 0,5k_p(x - x_d)^2 \tag{5.2}$$

where k_p is the coefficient influencing the field's attractive force; x, current vehicle's coordinate; and x_d, goal's coordinate.

If we rewrite the expression (5.1) in the form of $f(x) = 0$ and substitute the current vehicle's coordinates into it, we can determine whether the obstacle touches the vehicle or not.

For the given shortest distance ρ_0 between the obstacle and the vehicle and for the known current distance ρ, the repulsive force will have the following form

$$U_0(x) = \begin{cases} 0,5\eta\left(\dfrac{1}{f(x)} - \dfrac{1}{f(x_0)}\right)^2, & \text{if } f(x) \leq f(x_0) \\ 0, & \text{if } f(x) > f(x_0) \end{cases} \tag{5.3}$$

where η is the coefficient influencing the field's repulsive force.

In a more general case, the repulsive force will have the following form

$$U_0(x) = \begin{cases} 0,5\eta\left(\dfrac{1}{\rho} - \dfrac{1}{\rho_0}\right)^2, & \text{if } \rho \leq \rho_0 \\ 0, & \text{if } \rho > \rho_0 \end{cases} \tag{5.4}$$

The proposed method allows us to distribute the motion planning task between the path planning level and the lower one—the drives control level. The motion equations are determined through a Lagrangian accounting for Coriolis and centrifugal forces.

The acceleration of the manipulator's end point is composed of the components necessary for goal reaching, gravity compensation, and obstacle avoidance. It is found as a field gradient. The stabilizing control is formed using a proportional-differential law accounting for speed limitations put on the vehicle.

The force determining the motion direction of an arbitrary point of the manipulating robot in the repulsive field of an obstacle can be calculated as follows

$$
F^*_{(O,\text{psp})} = \begin{cases} 0,5\eta \left(\dfrac{1}{\rho} - \dfrac{1}{\rho_0} \right) \dfrac{1}{\rho^2} \dfrac{\rho}{x}, & \text{if } \rho \le \rho_0 \\ 0, & \text{if } \rho > \rho_0 \end{cases}.
\tag{5.5}
$$

Control for any point of a vehicle is a sum of controls generated for all the obstacle points. The motion limitations are set in the form of repulsive fields of the obstacle's bounding points.

The approach presented in Ref. [8] reduces the computational load at the top control level. This speeds up the vehicle's response as a reaction to the environmental changes. The form of the repulsive function is simple and does not require calculation of distances. Knowing the full field picture, a safe path can be planned.

However, it should be mentioned that the surface presented in Eq. (5.1) is hard to use for asymmetric obstacles. This narrows the application areas of the approach in Ref. [8]. The repulsive functions in Eq. (5.4) leave a space for local minima the vehicle can get into. Therefore, it is necessary to introduce special procedures to get out of them.

The work [9] considers the motion of a vehicle along a previously planned path. When an obstacle is reached, the vehicle's control system determines the form and position of the obstacle and corrects the initial path. The obstacle avoidance trajectory is recalculated as the new data about the obstacles is received. The obstacle is represented by a set of points on the surface in a vehicle-fixed coordinate system. The obstacle form identification is not performed.

In this work the potential-fields method is based on creation of a coordinate grid defining the motion surface and including the obstacles and vehicle's motion path. The obstacles create forces repelling the vehicle, while the preset path creates attractive forces.

If there are no obstacles, the potential-field minimum is located on a preset path. The obstacles' fields shift this minimum.

In each grid cell the attractive potential field is determined by the expression

$$P_R = z(x, y)$$

$$= \frac{|y - a(x, y)|}{2} [m_2 (\text{sgn}(y - a(x, y)) + 1) - m_1 (\text{sgn}(y - a(x, y)) - 1)]$$

$$(5.6)$$

where P_R is the field's characteristic; $a(x, y) = \frac{X_0 - A(y) - x \sin \psi}{\cos \psi}$; $A(y)$, desirable vehicle's position in the stationary coordinates system; (X_0, Y_0), current vehicle's position in the stationary coordinate system; (x, y), point coordinates on a path in a fixed system of coordinates; $m_1 = 1/[L - A(y)]$, $m_2 = 1/A(y)$; and L, width of vehicle's working area.

The repulsive potential field of an obstacle is created in the same grid as the attraction field of the path. The repulsive field of an obstacle is a sum of the potential fields of the obstacle's individual points:

$$P_0 = \sum_{i=1}^{k} e^{-\frac{c_1(y - y_i)^2 + c_2(x - x_i)^2}{2\sigma}} \qquad (5.7)$$

where c_1, c_2, σ are the parameters depending on drive's dynamics and relative speed between the obstacle and the vehicle and (x_i, y_i), coordinates of ith point of the obstacle.

The function for the total potential field is given by

$$P_u = P_R + P_0. \qquad (5.8)$$

The motion trajectory is set in the form of a series of points $\Gamma = \{(x_i, y_i^{\min})\} | i = \overline{1, n}$, where n is the number of grid cells along the x-axis of the fixed system of coordinates.

The coordinate x_i of the vehicle's motion trajectory changes from 1 to n. The coordinate $y_i^{\min}$ is given by the expression $y_i^{\min} = \arg \min_y P_u(x_i, y)$ for $y \in \{y_{1...m} | y_{i-1}^{\min} - I < y_{1...m} < y_{i-1}^{\min} + I\}$, where I is value of the minimal change of trajectory with a minimal potential. Usage of the value I smoothens out the possible path gaps due to sudden displacements of the potential-field minimums. The potential minimum area is limited by the interval $y_i \in (y_{i-1}^{\min} - I, y_{i-1}^{\min} + I)$.

A smooth path is created section by section using only a part of obtained set of points with a minimal potential. This is explained by the fact that in the process of obstacle avoidance additional information can come out requiring further path adjustments.

The path is smoothed out in the following way: from the obtained series of points we select first k points (the value k is selected accounting for the possibility of further corrections of points positions). A medium point with coordinates $(x_{\text{mid}}, y_{\text{mid}})$ is found in this set and a motion direction through this medium and kth point is determined. Joining the initial position, medium and kth point, and knowing motion direction and calculated curvature coefficient, we get a path in the form of a spline

$$y(x) = a_5 x^5 + a_4 x^4 + \ldots + a_0 \tag{5.9}$$

where coefficients $a_0, a_1, \ldots, a_5$ are found by solution of the equations obtained from boundary conditions: $y(0) = 0$, $\dot{y}(0) = 0$, $\ddot{y}(0) = 0$, $y(x_{\text{mid}}) = y_{\text{mid}}$, $\dot{y}(x_{\text{mid}}) = t$, and $\ddot{y}(x_{\text{mid}}) = 0$.

Based on the expressions (5.6)–(5.9), the following step-by-step path planning algorithm is proposed.

Step 1. The desired trajectory is set according to the general motion map without obstacles.

Step 2. The environment is scanned. If no obstacles are found, the vehicle starts motion along the preset path.

Step 3. If obstacles are found:

    3.1. Coefficients c_1, c_2, σ are tuned. A function describing the potential field (Eq. 5.8) is calculated.

    3.2. Potential function minimum points are calculated.

    3.3. The first k reliable points are selected and the next path section is calculated.

    3.4. Detour path is followed.

    3.5. If the built detour path does not contradict the sensor system data, go to step 2, else to step 3.6.

    3.6. If the detour path becomes unsafe, the parameter k is adjusted and a jump to 3.1 is performed.

In the work Ref. [9] the limitations on the vehicle's motion are not considered.

The obtained potential-field function has a simple form and is calculated in real time based on sensor system information and environment map. The function form leaves place for adding additional information concerning the vehicle's and obstacles' relative speed in the space of working coordinates.

A drawback of this approach is that it is assumed that we know the environment's map when planning the vehicle's path. The path's smoothness and computational speed depend on the selected coordinate grid cell size.

In work Ref. [10] it is proposed to change the safety circle for avoiding mobile obstacles using the method of "collision cone." This cone is formed by the vehicle's speed vectors pointing to the direction of a possible collision at any moment. The change of the safety circle in the process of motion is performed so that the relative speed becomes directed outside the "collision cone." The potential field is changed in the motion process.

The considered approach is a representation of the artificial potential field in the phase space of external coordinates. The total potential field is set in the form of following expression

$$U = \frac{\Re}{\Re + D} \ln\left(\frac{\Re}{\delta_1}\right) - \ln\left(\frac{1}{\delta_0}\right) \tag{5.10}$$

where $\Re$ is the obstacle's safety circle radius; D, distance between the goal and the obstacle; δ_0 distance between the vehicle and the goal; δ_1, distance between the vehicle and the obstacle; for $V_R < 0$: $\Re = \min\left(\overline{\Re}, R + \frac{h}{r\beta}\right)$, $\beta = \frac{|V_\theta|}{\sqrt{V_\theta^2 + V_r^2}}$; for $V_R \geq 0$: $\Re = R$; h the settings parameter.

In case of several obstacles, a field of an obstacle with a minimal difference between δ_1 and $\Re$ is selected to be an active potential field. In order to avoid frequent switching of an active potential field for several close values of $\delta_{1i} - \Re_i$, the resultant field is determined as a weighted average of several fields.

The considered method allows us to plan the vehicle's motion speed. The vehicle is represented as a point Π, moving with a speed V_R, and the obstacle as a point O, moving with a speed V_0, with a safety circle R that includes the vehicle's safety circle τ as shown in Fig. 5.1.

The distance between the vehicle and the obstacle is r. The vector of relative speed between the vehicle and the obstacle is split into two components:

$$V_r = V_0 \cos(\beta - \theta) - V_r \cos(\alpha - \theta), \tag{5.11}$$

$$V_\Theta = V_0 \sin(\beta - \theta) - V_r \sin(\alpha - \theta). \tag{5.12}$$

If we assume that vehicle and obstacle are moving with constant speeds and $V_R < 0$, and if the following condition is satisfied

$$V_\theta^2(t_0) \leq \frac{R^2}{r^2(t_0) - R^2} V_r^2(t_0) \tag{5.13}$$

the vehicle motion speed vector with respect to the obstacle belongs to the set of the "collision cone."

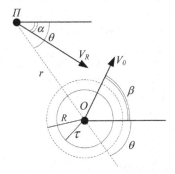

FIGURE 5.1 Mutual position of the vehicle and obstacle.

A vector determines the motion direction $E(p) = -\nabla U(p)$, where $U(p)$ is a potential-field function, p, position of the vehicle's center. The desired orientation direction of the vehicle is given by

$$\phi_d(p) = \arctg \frac{E_y(p)}{E_x(p)} \tag{5.14}$$

The desired vehicle speed can be found as

$$v_d(p, t) = \min\left(a_0 t, v_0, \sqrt{2a_0 \zeta(p)}\right) \tag{5.15}$$

where a_0 is the maximal acceleration; v_0, ground speed; and $\zeta(p) = \|p(t) - p_G\|$, distance to the goal point p_G.

For the large values of ground speed, the vehicle can get into the obstacle's safety circle at turning points. To avoid this event, the value of the ground speed is calculated according to the following equation

$$\dot{v}_0(p, t) = a_0 - v_0^2(p, t)\left\|\frac{\partial}{\partial p}\left(\frac{E(p)}{\|E(p)\|}\right)\right\| \tag{5.16}$$

Eq. (5.16) ensures reduction of the ground speed at the areas of large values of the potential-field gradient. In this approach of vehicle control, a relay controller is used with two sliding surfaces.

An advantage of this approach is that the exponential potential fields allow avoiding the local minima because the gradient lines start and end at the obstacles, the goal point, or infinity.

A drawback of this method is that it uses logarithmic functions that change their signs at certain distances to the obstacles. This can lead to undesirable change of motion direction while the authors state that such a situation is impossible due to the inherent system qualities.

In Ref. [11] virtual points are introduced in addition to the main goal point. They are located in the vicinity of the main point and the distance between these points and the goal point can be fixed or changing as a functioning result of a genetic algorithm searching for the motion path.

The path is formed to give a minimum for a criterion that includes distance to the goal and to the closest obstacle in each path point. Virtual goal points are used to get out of local minima, and the vehicle continues motion toward the goal after passing them.

The work Ref. [11] considers setting the fields using the methods of O. Khatib [8], Ge and Cui [12]. The approaches using Euclidian distances, exponential and logarithmic functions, and physical fields (e.g., temperature field) for setting potential fields are considered. The function of attraction to the goal points has the following form:

$$F_{att}(q) = -\nabla U_{att} = -\xi(q - q_a)\frac{1}{|q - q_a|} \tag{5.17}$$

where q is the vehicle coordinates; q_a, goal coordinates; and ξ, adjustment parameter.

Normalization makes the value of attraction function to be independent of the distance to the goal. The artificial repelling force is given by

$$F_{rep}(q) = \nabla U_{rep} = \begin{cases} \eta \cdot \sqrt{\dfrac{1}{d} - \dfrac{1}{d_0}} \dfrac{(q - q_o)}{d^3}, & d \leq d_0 \\ 0, & d > d_0 \end{cases} \qquad (5.18)$$

where q_0 is the obstacle coordinates; d_0, radius of the obstacle influence area; d, distance to the obstacle; and η, adjustment parameter.

Using the potential-fields function with a fast-growing amplitude allows the vehicle to pass through the narrow passages rather effectively. Square root of the potential-field functions creates smooth motion paths because such type of functions have higher amplitudes at average distances to the obstacles.

The adjustment parameters ξ and η are calculated using the optimization procedure based on a genetic algorithm that is also used to determine the coordinates of virtual goals.

When the approach is used to describe obstacles, it is sufficient to use coordinates of obstacle centers and influence radius. If the environment map is known, the obstacles are represented as polygons with set vertices and dimensions extended according to the vehicle size allowing us to represent it as a point.

For the regions without obstacles, a coordinate grid is introduced. A repelling force vector is calculated for each cell, and the attracting force vector is obtained afterward. These two vectors are then summed.

If the environment map is a priori unknown, the motion trajectory is calculated according to the optimization function. For the cells surrounding the vehicle, the optimization function value is calculated by the following formula

$$f(q) = \begin{cases} \sqrt{E/2} \cdot e^{-d_{min}}, & d_{min} > 0 \\ 2000, & d_{min} = 0 \end{cases} \qquad (5.19)$$

where d_{min} is the distance to the closest cell occupied with an obstacle; $E = |q_{rx} - q_{gx}| + |g_{ry} - g_{gy}|$; q_{rx}, q_{ry}, coordinates of the new possible vehicle position cell; and q_{gx}, q_{gy}, goal coordinates.

When the d_{min} value is calculated, the information about the known obstacles is used together with the sensor data covering the area of 5×5 cells. Selection among the surrounding cells is performed using the genetic algorithm based on a probabilistic approach.

Two variants are proposed as a solution of local minima problem. The first one is described in this section and it uses the creation of virtual goals at the polygon vertices representing obstacles. The second one is the usage of a hybrid approach. When the planner identifies getting into the local minimum, a switch to situational control is performed. In situational control, a vehicle motion mode is selected to ensure successful obstacle avoidance. After the situational control mode is completed, the control system switches back to the motion along the path calculated using the potential-fields method.

The work Ref. [11] presents the modeling results for a flat map including 40×40 cells with 5 randomly located obstacles. The authors use two virtual goal points to get the vehicle out of the local minima. Comparison of the two approaches (with set virtual goal points and virtual goal points found by the genetic algorithm) showed that the set location of the virtual points enables obtaining an optimal trajectory but can lead to collisions.

There are several advantages of the approach considered in Ref. [11]. The qualities of the obstacles' repulsive fields are improved. There is a possibility to avoid moving obstacles and virtual goal points can be controlled dynamically in order to avoid complicated obstacles.

At the same time, there are difficulties in passing through narrow passages. Auxiliary virtual goal points are necessary to pass obstacles of complicated configurations. The necessity of additional genetic algorithms and situation control algorithms increases the amount of time required for path planning.

In Ref. [13] a 2D coordinate grid is used for vehicle control. A value is assigned to each cell reflecting the probability of an obstacle being present in this cell. A vehicle is located inside a square of active cells (scanned by the vehicle's sensor system). Each active cell acts on the vehicle with a repulsive force $F_{i,j}$.

The expression for the repulsive function of an active cell has the following form:

$$F_{i,j} = \frac{F_{cr} W^n C_{i,j}}{d^n(i,j)} \left(\frac{x_i - x_0}{d(i,j)} \widehat{x} + \frac{y_j - y_0}{d(i,j)} \widehat{y} \right) \tag{5.20}$$

where F_{cr} is a coefficient of the repulsive force; $d(i,j)$, distance between the active cell (i,j) and the vehicle; $C_{i,j}$, the degree of confidence in placing obstacles in the active cell; W, vehicle's width; (x_0, y_0), current vehicle coordinates; and (x_i, y_j), coordinates of the active cell.

The resulting repelling force is given by the following formula

$$F_r = \sum_{i,j} F_{i,j} \tag{5.21}$$

The function of attraction to the goal is given by

$$F_t = F_{ct} \left(\frac{x_t - x_0}{d_t} \widehat{x} + \frac{y_t - y_0}{d_t} \widehat{y} \right) \tag{5.22}$$

where F_{ct} is a constant of attracting force; d_t, distance between the vehicle and the goal; and (x_t, y_t), goal coordinates.

In this approach the vehicle stability reduces at high speeds. The method proved its effectiveness in a series of experiments on real vehicles located in the environment with a known obstacles arrangement.

The method considered in Ref. [13] is rather easy to implement. However, it allows for getting into local minima and for oscillatory motion close to the obstacles and in narrow passages.

In Ref. [14] the authors consider the application of artificial potential fields in car race simulators. Using this method the vehicle's motion is planned along the track of an arbitrary form without obstacle avoidance. For implementation of the planning method a combination of multiagent technology and potential fields is used.

Multiagent technology is applied for simultaneous monitoring of a multitude of parameters, describing the current and predicted situations. This is necessary for online functioning of the planning system. Three types of fields are being used: shortest path field, track field, and track curvature field. Track is a coordinate space holding the vehicle's motion path. Six agents are used in this approach: driver, shortest path, track curvature, track, navigator, and interface. The "driver" is the main agent in the multiagent system structure. At each solution step, it determines a series of point-candidates for the possible motion located in various directions and distances from the vehicle.

A point with a highest potential is selected out of all the multitude of points.

$$p_{\max} = \max_{l \in L} \sum_{pf \in PF} \sum_{c \in pf} f_{pf}(\mathrm{dist}(l, c), \mathrm{size}(c)) \qquad (5.23)$$

where dist(l,c) is the Euclidian distance between the point l and location of the field source c.

The function f_{pf} determines the method estimating the influence of a certain field source. Once we find the point with a maximal potential, the agent "interface" sends an action necessary for motion to this point to the game server.

The agent "shortest path" tries to move the car along the shortest path on the track. A direction is selected for the car to move along the path of the shortest length. At the end of the straight segments, the attractive field sources are located. They are described by the following expression

$$f_{F \circ D}(d, c) = -kd^2 + c, \qquad (5.24)$$

where k is a preset constant; d, distance to the field source; and c, field source influence degree.

Before reaching a turn, the agent "track curvature" tries to move the vehicle so that the turn becomes less sharp maximizing the speed. The field source attraction function adjusting the turn trajectory has the following form:

$$f_{F \circ D}(d, c) = c(1 - d/d_{\max}) \qquad (5.25)$$

where $d_{\max}$ is the track width.

The field's influence degree is calculated using the following expression

$$c = \left(- 0,0002 msd_0^2 + 0,06 msd_0 - 2,5 \right) \cdot v_t \qquad (5.26)$$

where v_t is the current speed and msd_0, maximal visible distance.

The agent "track" is intended to hold the vehicle inside the track. For this purpose, a function of repulsion from the track boundaries is introduced in the following form:

$$f_{FoT}(d,c) = c \Big/ \sqrt{d} \qquad (5.27)$$

where c is the adjustment parameter.

During the test lap, the agent "navigator" determines the direction of turns on the track and then sends this information to the agent "track curvature" during the motion.

In Ref. [12] the repulsive function is formed in a way that accounts for the distance between the vehicle and the goal. Such a function ensures the location of the global minimum at the goal point. It should be mentioned that unlike many of other researches, the authors give the expressions for calculating the coefficients of the attractive and repulsive functions ensuring getting the vehicle out of the local minima.

A conventional function is used for the goal attraction function:

$$U_{att}(q) = \frac{1}{2}\xi\rho^m\left(q, q_{goal}\right) \qquad (5.28)$$

where ξ is the positive constant and $\rho^m(q,q_{goal}) = ||q_{goal} - q||$, distance between the vehicle and the goal $\left(m = \overline{1,2}\right)$.

The repulsive field function has the following form:

$$U_{rep}(q) = \begin{cases} \dfrac{1}{2}\eta\left(\dfrac{1}{\rho(q, q_{obs}) + \in} - \dfrac{1}{\rho_0}\right)^2 \rho^n(q, q_{goal}), & if\ \rho(q, q_{obs}) \le \rho_0 \\ 0, & if\ \rho(q, q_{obs}) > \rho_0 \end{cases}$$

$$(5.29)$$

where η is the positive constant; $\rho(q,q_{obs})$, distance between the vehicle; ρ_0, obstacle's influence radius; n, positive number; and μ, small constant.

In this work the functions coefficients were calculated for the cases when the vehicle is not in a line with the goal and the obstacle, and does not get into a local minimum. A situation is also considered when the goal is located between the vehicle and the obstacle. It should be mentioned that the derivation procedure for calculation of coefficients is described in detail, thus, if necessary, it is possible to derive the expressions for the case when the obstacle is located between the vehicle and the goal.

The vehicle does not get into a local minimum if the following conditions are satisfied

$$\frac{\xi}{\eta} > k_n \tag{5.30}$$

$$k_n = \sup_{\rho_m < \rho < \rho_0} \left\{ \left(\frac{1}{\rho} - \frac{1}{\rho_0} \right) (p - r)^{n-2} \alpha(\rho) \right\} \tag{5.31}$$

where r is distance between the obstacle and the goal; $\rho = \rho(q, q_{obs})$, $r < \rho < \rho_0$,
$$\rho_m = \frac{2r}{1 - \frac{n}{2} + \sqrt{\left(1 - \frac{n}{2}\right)^2 + \frac{2nr}{\rho_0}}} > r, \alpha(\rho) = \frac{1}{\rho} - \frac{r}{\rho^2} - \frac{n}{2\rho} + \frac{n}{2\rho_0} > 0 \text{ for } \rho > \rho_m.$$

In Ref. [15] there is an exact expression for k_n if $n = 2$, and a simplified one for $n \neq 2$. The obstacles in this work are presented by points with a certain influence radius. The vehicle's motion is described by the following equation

$$M(q)\ddot{q} + B(q, \dot{q}) + G(q) = \tau \tag{5.32}$$

where $M(q)$, vehicle's inertia matrix; $B(q, \dot{q})$, friction forces; $G(q)$, gravitational forces; and τ, control.

The feedback-based control is described by the following expression

$$\tau = G(q) - \nabla U_{total} + d \tag{5.33}$$

where d is a dissipative external momentum at the system's input.

For the path to be smooth it is necessary to require n, $m \geq 2$. In this method it is assumed that the vehicle's speed is constant and the potential field determines only the motion direction.

Application of the approach given in Ref. [15] allows us to solve the task of vehicle's motion in the environments with simple obstacles located close to the goal point. Just as in Ref. [12] the authors obtained an analytical expression for calculation of specific values of field functions coefficients allowing the vehicle to avoid getting into the local minima. However, it is necessary to identify conditions under which these expressions can be applied. Otherwise the coefficients should be recalculated at each planning stage. In addition, the presented expressions cannot take the vehicle out of the local minima located between the obstacle and the vehicle and created by the complicated obstacles.

There is a definition given by Ref. [16] "... the methods create artificial potential functions in the robot's working area, such that the function's global minimum is located at the robot's goal position and the local maximums are positioned at the places where the obstacles are located. The robot is 'propelled' by an artificial force proportional to the gradient of the potential function at the robot's location making it move toward the goal position...".

The methods considered in Ref. [16] belong to the methods of local reactive off-road navigation for ground-based autonomous robots. The potential field is defined in a 2D coordinate space accounting for the vehicle trajectory curvature. Such a representation accounts for the limitations regarding vehicle's dynamics, environment parameters, and navigational conditions (positions of the intermediate goals, obstacles' positions, and desirable speed) in the form of potential functions. The maneuvers are planned according to vehicle's capabilities based on the potential-field gradient.

The potential function keeping the vehicle from rollover has the following form:

$$PF_r(k, \nu) = \begin{cases} K_r \left(1 - \dfrac{(k - k_{\max})^2}{(k_r(\nu) - k_{\max})^2}\right) & k_r < |k| < k_{\max} \\ 0 & 0 \le |k| < k_r \end{cases} \qquad (5.34)$$

where $k_{\max}$ is the maximal path curvature that the vehicle can follow; K_r, potential-field function coefficient; and $k_r(\nu)$, function limiting the path curvature that can be followed by the vehicle without rollover and its value is given by the following expression:

$$k_r(\nu) = \frac{dg_z \pm hg_x}{h\nu^2} - \delta_r \qquad (5.35)$$

where d is the half of the vehicle's width; h, distance between the vehicle's mass center and ground surface; g_z, g_x, gravitational acceleration projections on the axes of the fixed system of coordinates; and δ_r, constant providing a stability margin while moving along a path with a maximal curvature.

A potential function limiting the vehicle's side slip has the following form

$$PF_s(k, \nu) = \begin{cases} K_s \left(1 - \dfrac{(k - k_{\max})^2}{(k_s(\nu) - k_{\max})^2}\right) & k_s < |k| < k_{\max} \\ 0 & 0 \le |k| < k_s \end{cases} \qquad (5.36)$$

where K_s is the potential function coefficient and $k_s(\nu)$, curvature function for a path that can be followed by the vehicle without a side slip given by the following expression:

$$k_s(\nu) = \frac{-g_x \pm \mu g_x}{\nu^2} - \delta_s \qquad (5.37)$$

where μ is the friction coefficient, and δ_s, constant providing a stability margin while moving along a path with a maximal curvature.

The potential functions of attraction to the intermediate goals are determined by the following expression

$$PF_w(k) = K_w(k - k_d)^2 \qquad (5.38)$$

where K_w is the potential function coefficient.

The potential function for reaching the desired speed is defined by the following expression

$$PF_v(k) = K_v(v - v_d)^3 \qquad (5.39)$$

where K_v is the potential-field function coefficient.

The potential function of repulsion from an obstacle is

$$PF_h(k, v) = \frac{K_h(K_{hv}v + 1)}{(K_{hd}O_d + 1)(K_{ha}|A_d| + 1)} e^{-(k-X)^2/2\sigma^2} \qquad (5.40)$$

where A_d is the a minimal angle between the next intermediate goal and an obstacle; $X = (k_1 + k_2)/2$, k_1 and k_2, maximal and minimal path curvature leading to the obstacle from the current vehicle position accounting for its speed; and $\sigma = (k_1 - k_2)/2$, K_h, K_{hv}, K_{hd}, K_{ha}, potential function coefficients.

The resultant potential-field function is given by the following expression

$$NPF(k, v) = PF_r(k, v) + PF_s(k, v) + PF_w(k) + PF_v(v) + \sum_{i=1}^{n} PF_{hi}(k, v),$$

$$(5.41)$$

where n is the number of obstacles.

Each item of the resultant potential function is added with a random noise keeping the vehicle out of the local minimums.

In Ref. [16] the vehicle's motion is described not in the Cartesian coordinates but in the phase space $\tau(k,v)$, where k, coefficient characterizing the path curvature; v, nonnegative speed of the vehicle's longitudinal motion along the path. The phase space allows forming the commands for the trajectory controller of lower level accounting for dynamic limitations.

Calculation of distance between the robot's gravity center and an intermediate goal creates a possibility to form a suboptimal trajectory with a minimal length performance index. A line connecting the vehicle's gravity center and an intermediate goal crosses the circumference with the radius of $2k_{max}$ centered at the robot's gravity center. The trajectory is selected so that the robot moves toward this intersection point. For the case when the distance between the vehicle's gravity center and the intermediate goal is less than $2k_{max}$ a path leading directly to the intermediate goal is selected.

The path curvature and speed are determined by $\tau^* = \tau + NF(\tau)$, where τ is the vehicle's position in the trajectorial space and $NF(\tau)$, resultant force of the potential field.

The gradient of the potential function is calculated as follows. The reachable rectangular area of the trajectorial space is split into nine similar rectangles with the center one corresponding to vehicle's current position. The potential function value is calculated for each of the rectangles. The direction of fastest descent is determined for the obtained surface. The desired vehicle's position is given by the rectangle (its central point) in the reachable trajectorial space.

The control law implementation accounts for the dynamic limitations put on the vehicle. The reachable speed ranges for the time interval t is given by the following equations

$$v^{max}_{reachable} = v + a^+ t \qquad (5.42)$$

$$v^{min}_{reachable} = v - a^- t \qquad (5.43)$$

where a^+, a^- are values of maximal acceleration and deceleration of the vehicle.

Maximal and minimal path curvatures are

$$\kappa^{max}_{reachable}(v) = k + \dot{k}_{max} t \qquad (5.44)$$

$$\kappa^{min}_{reachable}(v) = k - \dot{k}_{max} t \qquad (5.45)$$

where $\dot{k}_{max}$ is the maximal path curvature changing speed, while

$$\left| \dot{k}_{max} \right| = \frac{\tan \dot{\delta}_{max}}{L} \qquad (5.46)$$

where $\dot{\delta}_{max}$ is the vehicle's maximal rotation speed and L length of the wheelbase.

The modeling results presented in this work confirmed that application of the proposed approach allows the vehicle to avoid several obstacles and pass through several intermediate goals on its motion toward the final goal. The vehicle reduces the speed on sharp turns and the planner accounts for vehicle's dynamic qualities when moving at high speeds. The approach proved its effectiveness in modeling the motion on inclined surface. A series of successful experiments on real objects were performed where the task was to avoid one obstacle or visit two intermediate goals.

However, in Ref. [16] there are no mechanisms of getting out of local minima and the forces equations include many coefficients that have to be adjusted empirically.

The work [17] suggests using a random changing of repelling force to solve the local minima problem. It is proposed to use a virtual goal concept located so that the vehicle goes around a complicated obstacle and continues its motion toward the goal.

The attraction field is described by the following quadratic function

$$U_{att}(X) = \frac{1}{2}k(X - X_g)^2 \tag{5.47}$$

where X_g, is the goal position and k, attraction field function coefficient.

The repelling field is described by the following function:

$$U_{rep}(X) = \begin{cases} \frac{1}{2}\eta\left(\frac{1}{\rho} - \frac{1}{\rho_0}\right)(X - X_g)^2 & \rho \leq \rho_0 \\ 0 & \rho > \rho_0 \end{cases}. \tag{5.48}$$

Using the function (5.48) the location of the global minimum can be obtained at the goal point as in Ref. [12].

The resultant repelling force acting on the vehicle equal to negative gradient of the field function is determined by the following expressions

$$F_{rep} = \begin{cases} F_{rep1} + F_{rep2} & \rho \leq \rho_0 \\ 0 & \rho \leq \rho_0 \end{cases}. \tag{5.49}$$

As we can see from Fig. 5.2 the component F_{rep1} of the repelling force is directed away from the obstacle and F_{rep2} to the goal.

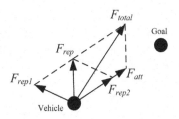

FIGURE 5.2 Components of the potential forces.

$$F_{rep1}(X) = \eta\left(\frac{1}{\rho} - \frac{1}{\rho_0}\right)\frac{1}{\rho^2}(X - X_g)^n, \quad F_{rep2}(X) = -\frac{n}{2}\eta\left(\frac{1}{\rho} - \frac{1}{\rho_0}\right)\frac{1}{\rho^2}(X - X_g)^{n-1}.$$

Function of the repelling force intended for getting the vehicle out of the local minima has the following form

$$F_{nrep} = AF_{rep} = \begin{cases} AF_{rep1} + AF_{rep2} & \rho \leq \rho_0 \\ 0 & \rho \leq \rho_0 \end{cases} \qquad (5.50)$$

$$\text{where } A = \begin{cases} \begin{bmatrix} \cos\alpha & -\sin\alpha \\ \sin\alpha & \cos\alpha \end{bmatrix} & \theta = \pi \\ \begin{bmatrix} 1 & 0 \\ 0 & 1 \end{bmatrix} & 0 \leq \theta \leq \dfrac{\pi}{2}, \end{cases}$$

θ, angle between the attracting force of the goal and the repelling force of the obstacle and α, random value in the range $0 < \alpha \leq \pi/2$. If the vehicle gets into a local minimum, a virtual goal is introduced close to the obstacle. When moving toward the virtual goal the real goal is ignored, and when the virtual goal is reached the attraction function becomes equal to zero. The force acting on the vehicle in the mode of motion toward the virtual goal is determined by the following expression

$$F_{rep} = \begin{cases} F_{rep1} + F_{rep2} + F_{vatt} & \rho \leq \rho_0 \\ 0 & \rho \leq \rho_0 \end{cases} \qquad (5.51)$$

where F_{vatt} is the attracting force of the virtual goal.

It is considered that the vehicle is stuck in a local minimum if the following condition is satisfied:

$$\sqrt{(x_{i+1} - x_i)^2 + (y_{i+1} - y_i)} < \varepsilon \qquad (5.52)$$

where ε is a small value set in advance.

The performance check of the proposed method demonstrated a successful avoidance of a rectangular and a U-shaped obstacle. This approach ensures reaching a global minimum and getting the vehicle out of the local minima. A drawback of the method is that using a random value in Eq. (5.50) leads to unexpected and jerky motions of the vehicle.

In Ref. [18] the vehicle's motion path is represented in the form of a virtual rubber band with its ends connected to the point of the vehicle's initial position and to the goal. The band is split into multitude of points, each forming its own attraction field. So the vehicle moves along its path being attracted to a succession of intermediate goal points. The coordinates x of the rubber band's points in the fixed coordinate system do not change. The y coordinates of the band's multitude of points are changed by the action of the potential fields created by the obstacles. This action deforms the trajectory.

The obstacles are set in the form of polygons while the diameter of the obstacles is increased to the vehicle's size allowing it to be represented as a point.

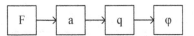

FIGURE 5.3 Vehicle coordinates and orientation angle calculation sequence.

Such an approach solves the problem of getting into local minima but leaves the question of selecting the positions of the intermediate goal points.

5.1.3 Potential-Fields Method Application Examples

A modeling of a vehicle's motion control using potential-fields method was performed using a calculation sequence for coordinates and orientation angle presented in Fig. 5.3.

In the beginning, according to Fig. 5.3, a resultant force F acting on the vehicle is calculated along with its projections F_x, F_y on the axes of the basic coordinate system. Then the vehicle's linear acceleration a is found using the Newton's second law. Then, the vehicle's position coordinates q are found in the basic coordinate system using the following expression

$$q(t) = q_0 + v_0(t) + \frac{at^2}{2} \tag{5.53}$$

and the orientation angle is determined

$$\phi = \arctan\left(\frac{a_{p_2}}{a_{p_1}}\right). \tag{5.54}$$

The motion planning under uncertainty conditions leads to some formalization of forces as shown in Fig. 5.4.

When a locator is used, the vehicle's field of view captures certain points of the obstacles (Fig. 5.4A). Thus, motion planning methods can account for the influence of all the visible points or of the point closest to the vehicle. In other words, using the potential-field method we can calculate the influence of each point of the obstacle and then seek for the resultant repelling force (Fig. 5.4B). Alternatively, the influence of a single closet point can also be used (Fig. 5.4C).

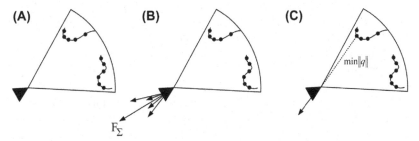

FIGURE 5.4 Specific formalization of forces acting on the vehicle.

Table 5.1 Potential Forces Used for Modeling

Attracting	Repelling
$F_{att}(q) = k\frac{(q_r - q_g)}{q_{rg}}$, (by Eq. 5.17)	$F_{rep}(q) = c\frac{(q_r - q_o)}{q_{ro}^2}$, (by Eq. 5.5)
	$F_{rep}(q) = F_{rep1}(q) + F_{rep2}(q)$, (by Eq. 4.9)
	$F_{rep1}(q) = c\left(\frac{1}{q_{ro}} - \frac{1}{r}\right)\frac{1}{q_{ro}^2}(q_r - q_g)^n$,
	$F_{rep2}(q) = -c\frac{n}{2}\left(\frac{1}{q_{ro}} - \frac{1}{r}\right)(q_r - q_g)^{n-1}$,
	$F_{repx} = \frac{1}{q_{ro}}\cos(\theta_{og})$, $\hspace{3cm}$ (5.55)
	$F_{repy} = \frac{1}{q_{ro}}\sin(\theta_{og})$,
	$F_{rep}(q) = AF_{rep1}(q) + AF_{rep2}(q)$, by (Eq. 5.50)
	$A = \begin{pmatrix} \cos(\alpha) & -\sin(\alpha) \\ \sin(\alpha) & \cos(\alpha) \end{pmatrix}$, with
	$\gamma = \delta\pi, \alpha = \text{rand}\left(0, \frac{\pi}{2}\right)$,
	$F_{repx} = \frac{1}{q_{ro}^2}\cos(\theta_{og})$, $\hspace{3cm}$ (5.56)
	$F_{repy} = \frac{1}{q_{ro}^2}\sin(\theta_{og})$,
	$F_{rep}(q) = c \cdot e^{-\beta q_{ro}}$ by (Eq. 5.7)

The modeling of potential-field method application results was performed with forces having analytical expressions presented in Table 5.1.

Table 5.1 uses the following denotations q_r, q_g, q_o vehicle coordinates, goal coordinates, and obstacle coordinates respectively; q_{ro}, q_{rg} distances between the vehicle and obstacle points, and between the vehicle and the goal, respectively; k, c, n, δ, β coefficients; r, obstacle's safety radius; θ_{og}, angle between the obstacle and the goal; γ, angle between the vehicle and the goal.

In all the experiments, a normalized attracting force was used to ensure the field's constant amplitude. The vehicle's motion modeling results with potential forces listed in Table 5.1 are presented in Fig. 5.5.

All the experiments except for the one with the functions given by Eq. (5.49) resulted in correct vehicle's motion in a "simple" scene. An incorrect motion in Fig. 5.5B can be explained by the fact that repelling potential functions include the difference between the distances to the obstacle and the goal that reduces the amplitude of the force as the vehicle approaches the goal (see Table 5.1, second row).

A change in coefficients of repelling forces influences the vehicle's motion safety and the whole path character as shown in Fig. 5.6 for the expression (5.5) with the coefficients reduced by the factor of 2.

Comparing the modeling results, we can conclude that most of the methods are acceptable for planning a vehicle's motion in an environment with simple obstacles.

Fig. 5.7 demonstrates that in all the experiments a correct motion is ensured for scene 2.

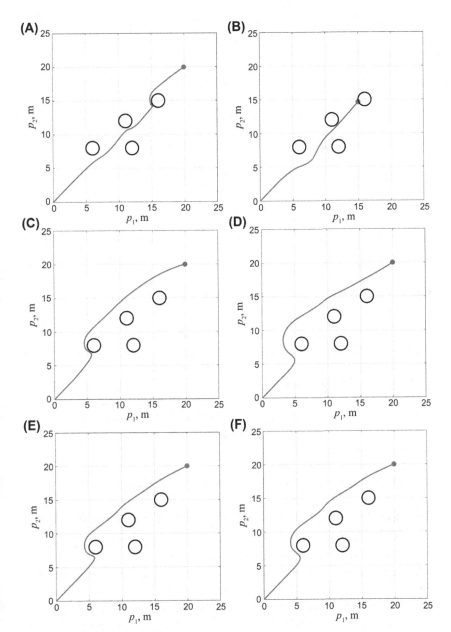

FIGURE 5.5 Scene 1 modeling results.

Fig. 5.8 shows that application of the function (5.49) leads to an incorrect motion in scene 3, i.e., the vehicle can get into local minima even with the obstacles of a simple shape.

Fig. 5.9 presents the modeling results for scene 4 demonstrating that in scene 4 only the function (5.55) ensures reaching the goal. In all the other cases, the motion quality was unsatisfactory.

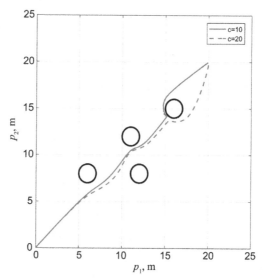

FIGURE 5.6 Modeling results with reduced coefficients of repelling forces.

Fig. 5.10 shows the modeling results for scene 5. Application of the functions (5.7), (5.50), and (5.55) makes it possible to reach the goal due to early maneuver. In all the other cases the motion was unsatisfactory.

Numeric values for the experiments are presented in Table 5.2.

Analysis of the obtained data leads to the conclusion that if the scenes become more complicated, the potential-field method would not guarantee reaching to the goal because there is no mechanism for getting out of local minima.

The simplicity of motion planning using the potential-field method makes it feasible to use it in more complicated intellectual algorithms.

5.2 APPLICATION OF VORONOI DIAGRAMS TO PATH PLANNING

Voronoi diagram is partitioning of a plane into regions (convex polygons called cells) based on the distance to a set of n points (called seeds) in such a way that for each seed there is a corresponding region consisting of all points closer to that seed than to any other. Consider a set of points S and a certain point p. For the point p, the Voronoi cell is an intersection of half-planes formed by central perpendiculars of the segments that include the point p. There are extensions where the point objects are represented by segments or curves. In vehicle motion planning, such objects are obstacles or their parts.

The path lengths obtained using Voronoi diagrams sometimes differ from the optimal ones. This is caused by the fact that the requirements of maximal distance to the obstacles and minimal path length are frequently contradictory.

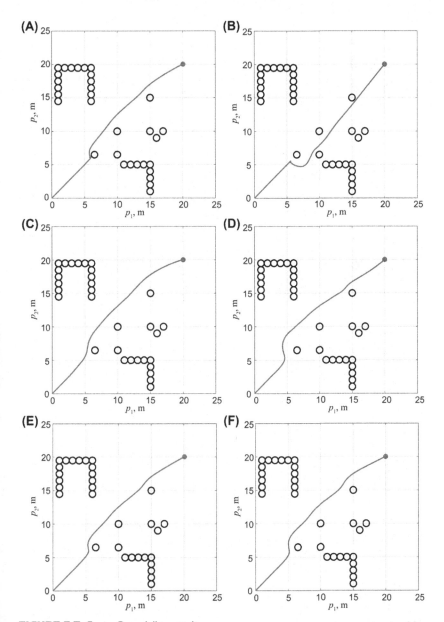

FIGURE 5.7 Scene 2 modeling results.

5.2.1 Survey of Voronoi Diagrams Applications Examples

One of the main approaches of vehicle path planning extensively uses the methods of computational geometry. The vehicles and the environment are described using various geometrical shapes. For example, if the vehicle has

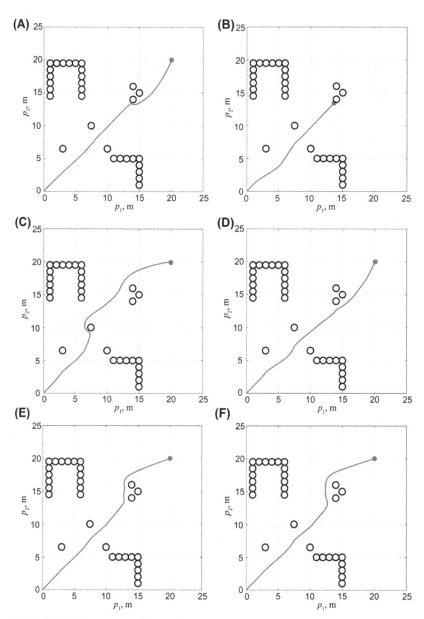

FIGURE 5.8 Scene 3 modeling results.

the form of a polygon, an approach based on Minkowski Sums is frequently used [19].

According to Ref. [20] the classical path planning algorithms can be split into two main groups: method of partitioning into cells [21] and the roadmap method [22].

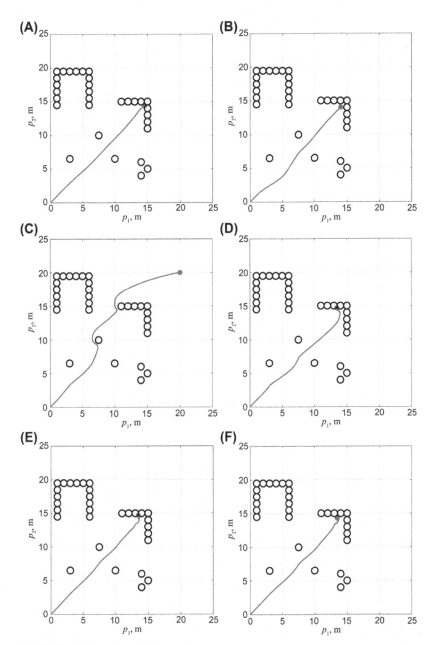

FIGURE 5.9 Scene 4 modeling results.

In the cell partitioning method, the space C_f free for motion is described by nonoverlapping cells. The partitioning can be either exact or approximate. In the approximate partitioning, the space is being split until each cell is either completely in the free space C_f or complete inside an obstacle. The process of recursive partitioning is stopped when a preset accuracy is reached. The exact

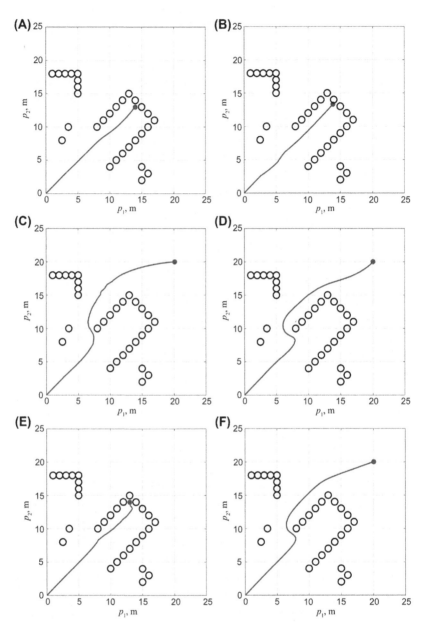

FIGURE 5.10 Scene 5 modeling results.

cell partitioning works faster than the approximate one but the obtained paths become longer.

In the roadmap method [22], the free space connections are described by a graph. This method has several variations. The most widely spread one is the probability based roadmap method [23]. The other popular methods are based

Table 5.2 Experimental Values of Motion Criteria

Criterion	Example 1	2	3	4	5	6
Safety criterion S_m	0.15	0	0.20	1.28	0.38	0.51
	0.19	0.14	0.68	0.80	0.65	0.91
	0.18	0	0.23	0.68	0.69	0.92
	0	0	0.23	0	0	0
	0	0	0.51	0.86	0	0.54
Path length P_L	29.73	∞	31.55	33.23	31.24	32.40
	31.40	30.09	29.52	29.73	29.01	29.35
	30.05	∞	31.05	28.70	29.85	30.40
	∞	∞	31.91	∞	∞	∞
	∞	∞	32.11	32.10	∞	31.54
Task completion time t_m	36.80	∞	39.00	41.50	38.80	40.30
	39.10	37.60	36.00	36.70	35.30	35.70
	37.10	∞	38.20	35.00	36.30	37.50
	∞	∞	39.70	∞	∞	∞
	∞	∞	40.20	40.10	∞	38.80
Mission success coefficient F	1	0	1	1	1	1
	1	1	1	1	1	1
	1	0	1	1	1	1
	0	0	1	0	0	0
	0	0	1	1	0	1

on such structures as a visibility graph in searching for the shortest path and Voronoi diagram in searching for a free way.

Comparing the computational complexity of the last two algorithms, we can conclude that Voronoi diagram is built in $O(n\log n)$ clock cycles for n nodes of the network, while the visibility graph is built in less than $O(n^2)$ cycles for the same map.

The mentioned drawback can be avoided using special path smoothing methods [24]. In Ref. [25] the authors have combined the Voronoi diagram with the visibility graph and potential-field method to attain a compromise between the safest and the shortest path. Though the obtained path is shorter than the one obtained using the Voronoi diagram, it is not optimal.

Since path planning using Voronoi diagram with a multitude of polygons is a complicated and time-consuming process, the authors of Ref. [19] have

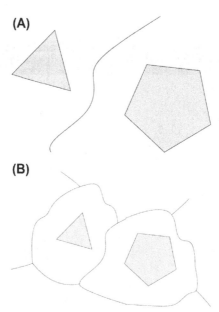

FIGURE 5.11 Full (A) and incomplete (B) roadmaps.

approximated the polygonal obstacles by points on their edges and built diagrams using these points. The edges that were intersecting the obstacles were deleted. First, a Delaunay triangulation was performed. Later a Voronoi diagram was created in $O(n)$ clock cycles.

A random incremental algorithm was used to create the Delaunay triangulation. It allows adding new points without changing all the triangulation.

The process of building begins with creation of a triangle including all the approximating points. The corner points of this triangle should not be located inside the circles describing the internal triangulation triangles. Once the large top triangle is built, the points are added one by one. The necessary topological changes are performed with addition of the points to retain the qualities of Delaunay triangulation.

In Ref. [19] the next step involves building a roadmap where the triangulation is added with a starting point and a goal point. The roadmap is generated by removing of Voronoi diagram edges that are located closer than the doubled safety radius ($2C_{min}$). However, if the Voronoi diagram includes only the obstacle points, the roadmap will be incomplete.

Fig. 5.11A and B shows incomplete and full roadmaps, respectively.

A full roadmap can be obtained by building a covering frame around the obstacle vertices and extending it by not less than $2C_{min}$ in all directions.

A Voronoi diagram is then built by points approximating obstacles and by surrounding frames. Based on this diagram, a full roadmap is obtained as shown in Fig. 5.11B.

The shortest path in the resulting roadmap can also essentially differ from the optimal one due to a convex shape of the obstacles. To solve this problem, the approximating points of the covering frame are pinched to a closest obstacle point to a distance not less than $2C_{min}$ [19]. The Voronoi diagram is built using the approximating points of the obstacle and the points on the frame. If it becomes necessary to set new starting and goal points, the previous points are removed from triangulation dynamically.

In Ref. [19] the authors have proposed a simple approach reducing the length of the shortest path on the roadmap. For the vertex v_i ($i = \{1, \ldots, n-2\}$) it is checked if the distance to the closest obstacle of the segment $\overline{v_i v_{i+2}} \leq C_{min}$. If this condition is satisfied, the vertex v_{i+1} is deleted from the shortest path and the search process continues. After the path shortening procedure is completed, the path is smoothed using the Steiner points [26].

In general, the proposed approach yields a path optimal in the sense of safety and length. However, due to the presence of several iterational algorithms, it has a low computational effectiveness and the logic behind selection of the parameter C_{min} is not obvious.

In Ref. [27] a boundary-widening algorithm is used to build Voronoi diagrams. Although this work is devoted to map analysis and not to the motion toward the goal, some of the approaches discussed can be used for successful obstacle avoidance.

In Ref. [27] the authors introduce a notion of an intersection point P where the following conditions are satisfied: (1) there is a circle centered at the point P touching the obstacles' boundaries at more than two points called the *closest points* and (2) the interior of the circle does not intersect any obstacles. Such a circle with four points of contact is shown in Fig. 5.12. The lines connecting the intersection point P and the closest points on the obstacles' boundaries partition the intersection circle into sectors.

Suppose that the vehicle moves between the obstacles with the surfaces ∂O_i and ∂O_j along the Voronoi edge until it reaches an intersection point P. The

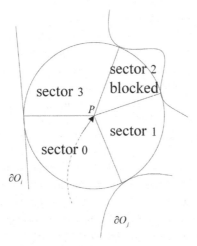

FIGURE 5.12 Intersection circle with 4 sectors.

sector through which the vehicle moves is defined as sector 0. All the other sectors are indexed in the counter-clockwise direction. When two end points of a particular sector are on the same obstacle, the sector is called a *blocked sector*, which is illustrated as "sector 2" in Fig. 5.12.

The rest of the sectors are called *open sectors*. If the intersection detected by the vehicle is adjacent to an open sector that has not been visited by the vehicle, then the intersection is marked as unexplored.

In the boundary expansion algorithm proposed by the authors of Ref. [27], it is assumed that the vehicle always passes the obstacles on the right and the planner can tell the difference between the ordinary obstacles and the ones that mark the environment boundaries.

The algorithm starts with a vehicle motion around the closest obstacle O_i. The vehicle tracks the Voronoi edges surrounding the obstacle moving in a clockwise direction. It records the visited intersection circles centered at P points.

If the vehicle completes a loop around an obstacle, its trajectory becomes a part of a Voronoi diagram called enclosing boundary of an obstacle O_i. After the loop is completed, a vehicle's trajectory is build going through unexplored sectors of the found intersection circles centered at P points.

So the path planner tries to form paths in a way to obtain the enclosing boundaries of the new obstacles. All the obtained enclosing boundaries of the obstacles are added to the global Voronoi diagram of the environment being explored. A full Voronoi diagram creation for the environment M with hexagonal Voronoi cells takes the following time

$$\tau = t\left(\frac{1}{2}M^2 - \frac{1}{2}M\right)$$

where t is the average time for a vehicle to traverse along the edges of one hexagonal Voronoi cell.

It should be mentioned that the presented approach ensures building a full map of the explored environment but does not guarantee correctness of marking sectors of intersection circles as blocked or open. For the terrain exploration task, a wrong marking of a sector as blocked does not have a significant importance, but in motion planning this can lead to unjustified path lengthening.

In Ref. [28], to explore flat and 3D environments, it is proposed to plan paths based on the Voronoi Fast Marching Method. This method is based on the Logarithm of the Extended Voronoi Transform (EVT) and the Fast Marching Method.

Logarithm of EVT models a repulsive field of the obstacles and the method of Fast Marching implements the wave propagation along the vector map of potentials. The path is calculated using the gradient method.

Application of the combination of approaches presented in Ref. [28] makes the vehicle move to the most unexplored free areas of the environment.

EVT is a convenient instrument for digital image processing. It calculates a Euclidian distance for a binary image. For each pixel in the image, the EVT assigns a number which is the distance to the nearest nonzero pixel of the image.

In a binary image, a pixel is referred to as background if its value is zero. For a given distance metric, the EVT of an image produces a distance map of the same size. For each pixel inside the objects in the binary image, the corresponding pixel in the distance map has a value equal to the minimum distance to the background. The EVT is closely related to the Voronoi diagram.

To put it another way, the Voronoi Fast Marching Method works by the principle of wave propagation in a heterogeneous environment. The wave propagation speed is calculated using the gradient method.

Application of the Fast Marching Method based on non-stationary fields essentially improves the obtained path quality. On one hand, the paths approach the Voronoi diagram edges; on the other hand, they are smooth. Since the paths are generated using the Logarithm of EVT, the potential-field functions have the shape that together with wave propagation principle allows us to solve the problem of local minimums.

There are methods that not simply combine the Voronoi diagrams with other methods but modify the Voronoi diagram itself, e.g., the mobile robot path planning method for a non-defined environment based on a generalized Voronoi graph [29].

5.2.2 Improvement of the Known Solutions by Mapping

The raw locator data require enormous volume of computations in order to build a Voronoi diagram. Indeed, each stage of the vehicle's motion is supported by a multitude of coordinates given by the locator. Without a serious processing these locators cannot tell us which point belongs to which obstacle; what is the orientation of the obstacles with respect to the vehicle and the goal; and if the obstacle coordinates can be connected.

For the effective motion planning under the conditions of uncertainty, mapping is to be performed and has to be synchronized with the vehicle's motion.

The easiest but the least effective way is to save all the locator data and to go through all the coordinates to analyze what obstacles they belong to. As the amount of information increases, the computational load makes it impossible to perform real-time mapping. The approach proposed in this work is illustrated in Fig. 5.13. The mapping process has the following stages:

Stage 1. Receiving locator data and joining it into clusters by analysis of distances between the points in the procedure of exhaustive search, and eliminating redundancy by deletion of point coordinates with a norm $\|r\| < \delta$ between the neighboring points (Fig. 5.14). This results in an array of cells with coordinates of objects in the vehicle's range of vision.

Stage 2. Merging the array of cells with object coordinates in the vehicle's range of vision with the base of cells with coordinates of all the known objects.

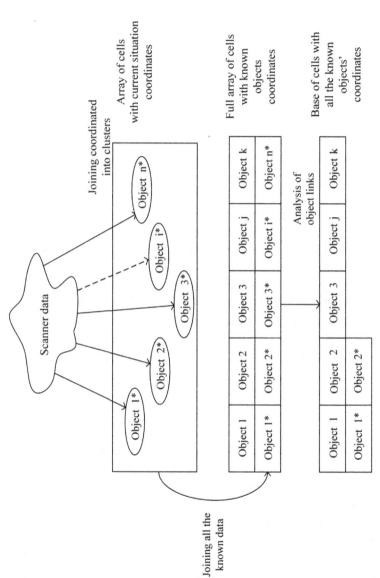

FIGURE 5.13 Mapping process.

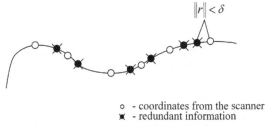

o - coordinates from the scanner
✖ - redundant information

FIGURE 5.14 Elimination of information redundancy.

Stage 3. Analysis of the links for merging the coordinate clusters belonging to a single or different objects located so that the vehicle cannot move in between.

Let us consider how the information coming from the locator is cohered with the existing information in the base of cells with coordinates of all the known objects (Fig. 5.15). In Fig. 5.15A, the coordinates of points obtained from the locator belonging to a single object intersect with the coordinates stored in the base.

In Fig. 5.15B, the new and known coordinates belonging to a single object do not intersect.

For reducing the computational load and eliminating an exhaustive search, stage 3 should be performed in two phases. For information merging in the first case (Fig. 5.15A), polygons are built for all the objects out of the full array of cells and their pairwise intersections are found. This can be done using the algorithms of Weiler-Atherton [30] or Bentley-Ottmann [31].

If a pair of polygons has at least one common point, the two coordinate clusters are joined. For merging the information in the second case (Fig. 5.15B), it is possible to use the locator's data representation pattern: in the array of coordinates the data is located from left to right. So in the array of cells with the current data, the coordinates of each object are stored from left to right. Therefore, it is sufficient to compare the norms of distances between the outermost coordinates in each pair of clusters and to determine if they belong to the same of different objects as it is illustrated in Fig. 5.16.

After the completion of the third stage, we get database with coordinates of all the known objects.

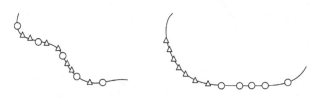

O - old coordinates
Δ - current scanner coordinates

FIGURE 5.15 Possible combination of old and new object coordinates.

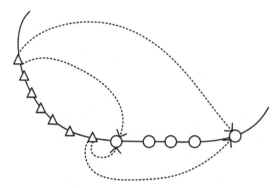

FIGURE 5.16 Analysis of the clusters of coordinates belonging to the same object.

Elimination of the exhaustive search of coordinates in large arrays of data makes this computational algorithm effective and suitable for real-time functioning in the vehicle's motion planning.

5.2.3 An Example of Planning Algorithm Implementation Using Voronoi Diagram

For implementation of the vehicle's motion planning algorithm, let us consider motion in the environment with obstacles shown in Fig. 5.17.

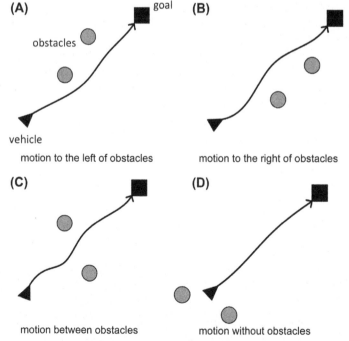

FIGURE 5.17 Motion modes in the environment with obstacles.

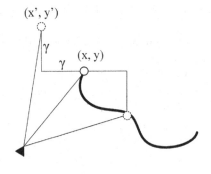

▲ - Vehicle

FIGURE 5.18 Motion to the left of the obstacle.

The mode of motion without obstacles is organized in case when there are no obstacles in the vehicle's range of vision. Motion to the goal in this case can be organized by setting the attracting potential to the goal.

Let us consider organization of motion to the left and right of the obstacles. After the obstacle nearest to the vehicle is determined, its nearest point is selected. Motion to the left of the obstacle is illustrated in Fig. 5.18. When ith obstacle is led to pass on the left, the coordinates (p_1, p_2) of the first point of the ith cluster are selected; if it passes on the right, the last point coordinates are selected. The coordinates increment is set as $(p_1 + \gamma, p_2 - \gamma)$ and $(p_1 - \gamma, p_2 + \gamma)$ (Fig. 5.18).

Out of these two points, the one which is most distant from the obstacle is selected (x', y') for the vehicle to move toward. This ensures a safe obstacle avoidance.

Fig. 5.19 demonstrates the mode of motion between the obstacles, i.e., steps $(K - 1)$, K, and $(K + 1)$. Motion in this mode can be organized as a succession of stages.

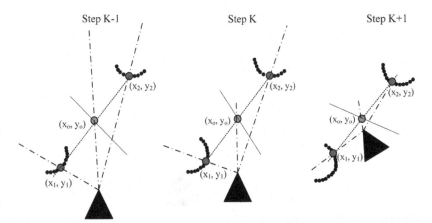

FIGURE 5.19 Motion between the obstacles.

Stage 1. Search for the coordinates of the obstacle points (p_{11}, p_{21}), nearest to the vehicle.

Stage 2. Search for the coordinates of the middle of the segment connecting the nearest points (p_{10}, p_{20}).

Stage 3. Motion to the point (p_{10}, p_{20}).

At each step, the point (p_{10}, p_{20}) belongs to a common edge of the Voronoi diagram polygons surrounding each of the obstacles. This corresponds to the case of incomplete roadmap, since the vehicle moves under the conditions of uncertainty. In order give smoothness to the path, it is possible to organize motion along the arc of a circle as it is shown in Fig. 5.20.

Motion along the arc consist of four stages.

Stage 1. Search for the coordinates of the nearest points (p_{11}, p_{21}), (p_{12}, p_{22}).

Stage 2. Search for the coordinates (p_{10}, p_{20}) of the middle of the segment connecting the nearest points and calculation of the radius R of the circle with its center located at the obstacle nearest to the vehicle.

Stage 3. Search for the coordinates (p_{1cir}, p_{2cir}) of the intersection of the circumference and the line connecting the vehicle with its center.

Stage 4. Motion to the point (p_{1cir}, p_{2cir}).

Effectiveness of the proposed algorithm in the mode of motion between the obstacles is confirmed by the modeling results presented in Fig. 5.21.

Fig. 5.21 shows the vehicle effectively avoids the obstacles. Additional modeling was performed for several scenes. The results are presented in Fig. 5.22.

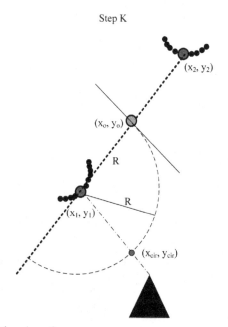

Step K

(x_2, y_2)

(x_o, y_o)

R

R

(x_1, y_1)

(x_{cir}, y_{cir})

FIGURE 5.20 Motion along the arc.

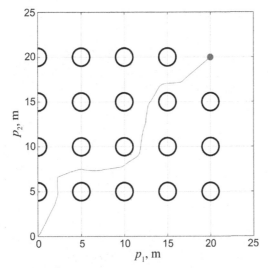

FIGURE 5.21 Motion along the obstacles modeling results.

From the modeling results, the vehicle failed to pass the scenes as shown in Fig. 5.22D and E. This is because the algorithm does not include the mode of motion along the obstacle. The values of performance indexes are presented in Table 5.3.

The method's effectiveness is estimated by the four criteria considered above.

The second column of the Table 5.3 presents the coefficient a_n and its value is a measure of importance of the corresponding criterion. The third column presents the five values corresponding to the five test scenes for each criterion. If a scene is failed the safety index in zeroed and the path length and completion time are assigned infinity values.

Analysis of the obtained data shows that effective vehicle motion planning methods using Voronoi diagram for a priori unknown obstacle positions require mapping to reduce the uncertainty.

5.3 VEHICLE MOTION PLANNING ACCOUNTING FOR THE VEHICLE'S INERTIA

5.3.1 General Notions

An essential drawback of many of the developed path planning methods is the absence of proofs of realizability of these paths accounting for the kinematic and dynamic qualities of the vehicle. For example, if the path makes a sharp turn, the vehicle moving at a high speed can hit the obstacle.

It is especially important to account for the kinematic and dynamic qualities when moving in unknown environments. Under such conditions the planner functioning bases mostly upon the information obtained from locators. However the locators have a limited range. As a result, the planner forms the path sections

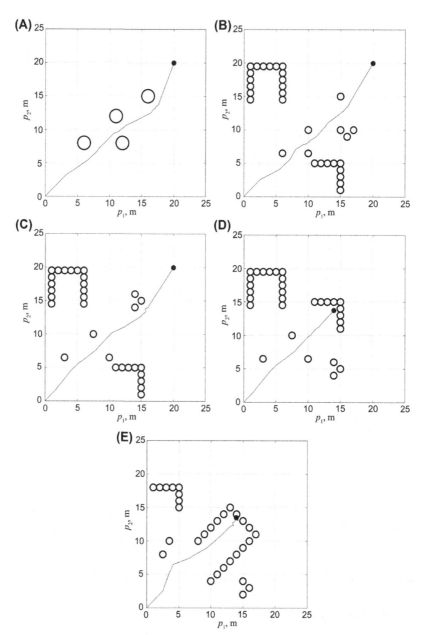

FIGURE 5.22 Vehicle motion modeling results.

of a limited length, and a vehicle with a limited maneuverability is not always capable of tracking paths with significant curvature.

Therefore, it is necessary to solve the path planning task not only in the co-ordinate space but also in the space of speeds, which allows us to form a feasible path.

Table 5.3 Values of Performance Indexes

Criterion	a_n	Planner Using Voronoi Diagrams
Safety criterion S_m	0.2	0.82
		0.60
		0.78
		0
		0
Path length P_L	0.2	29.41
		29.20
		29.13
		∞
		∞
Task completion time t_m	0.2	35.10
		34.80
		34.30
		∞
		∞
Mission success coefficient F	0.4	1
		1
		1
		0
		0

In obstacle avoidance, it is necessary to account for the three groups of vehicle parameters [32]: geometric parameters, its kinematics and dynamics. Geometric parameters of a vehicle and its kinematics are addressed together as a geometric task of representing the robot's position in the state of feasible paths. The vehicle's dynamics is determined by its maximal acceleration and deceleration. To ensure a safe motion it is necessary to select such a control that is realizable in a limited time interval for the known current speed and accounting for the braking length.

According to Ref. [32] at present the problem of obstacle avoidance accounting for the geometric, kinematic, and dynamic qualities of a vehicle is solved using the following three approaches:

1. development of methods to include the limitations into the planning methods [33];
2. development of methods allowing to abstract from limitations [34]; and

3. breakdown of a general task into subtasks and considering limitations after the planning procedure [35,36].

5.3.2 Survey of Known Approaches to Planning Accounting for the Inertia of the Vehicles

In the context of the approaches using the space of allowed speeds, the limitations on the kinematics and dynamics are explicitly accounted for. In such approaches the vehicle is supposed to move along the arcs. The main methods using the space of allowed speeds are: curvature-velocity method [37], lane-curvature method [38], and dynamic window approach [33].

The dynamic window approach described in [33] is based on using the space of allowed speeds for the search of feasible paths. The vehicle's dynamics is accounted by reducing the search space to admissible velocities that are reachable accounting for the introduced dynamic limitations. Additionally, in the search for the control actions, only the vehicle speeds that are considered to be safe are selected.

In Ref. [33] a path is considered as a group of n circular arcs called curvatures. Each curvature is defined by translational and rotational velocities (v_i, ω_i). To form the motion paths, the planner calculates the desired velocities (v_i, ω_i) for each arc. The search space grows exponentially with n. To reduce the computational load, the path is formed for the first of n arcs and the speed on the rest $n - 1$ arcs is assumed to be constant.

The speed is considered to be admissible if it allows the robot to stop safely. The set of V_a of admissible velocities is defined as

$$V_a = \left\{ (v, \omega) \middle| v \le \sqrt{2 \cdot \text{dist}(v, \omega) \cdot \dot{v}_b} \wedge \omega \le \sqrt{2 \cdot \text{dist}(v, \omega) \cdot \dot{\omega}_b} \right\} \quad (5.57)$$

where $\text{dist}(v, \omega)$ is the distance to the closest obstacle on the corresponding curvature, and v_b, ω_b the translational and rotational accelerations.

In order to take into account the limited accelerations, the overall search space is reduced to the dynamic window which contains only the velocities that can be reached within the next time interval t. The dynamic window is defined as

$$V_d = \{ (v, \omega) | v \in [v_a - \dot{v}t, v_a + \dot{v}t] \wedge \omega \in [\omega_a - \dot{\omega}t, \omega_a + \dot{\omega}t] \} \quad (5.58)$$

where (v_a, ω_a) are the current vehicle speeds, and $(\dot{v}, \dot{\omega})$ the accelerations.

The above given restrictions imposed on the search space for the velocities result in the area V_r defined as the intersection of the restricted areas

$$V_r = V_s \cap V_a \cap V_d \quad (5.59)$$

where V_s is the space of possible velocities.

In Ref. [33] the required velocity is selected from the V_r space to deliver maximum to the objective function

$$G(v, \omega) = \sigma(\alpha \cdot \text{heading}(v, \omega) + \beta \cdot \text{dist}(v, \omega) + \gamma \cdot \text{velocity}(v, \omega)) \quad (5.60)$$

where σ, α, β, γ are weighting parameters, heading (v,ω) the function measuring the alignment of the robot with the target direction, velocity (v,ω) the function used to evaluate the progress of the robot on the corresponding path—a projection of the translational velocity.

Generation of paths according to Eq. (5.60) provides motion to the goal with fast obstacle avoidance under the set dynamic limitations.

It should be mentioned that the objective function proposed in Ref. [33] does not exclude getting into the local minima area. In this case the authors propose to perform rotation away from the obstacle until the robot is able to translate again. For high-speed traveling through corridors and narrow doors, a speed-dependent side clearance is introduced—a safety margin that grows linearly with the robot's translational velocity.

In Ref. [39] an Improved Dynamic Window Approach is proposed that solves the problem of local minima. A velocity closest to the desired one is selected out of the set of admissible velocities. To find out this velocity, the following Lyapunov function is proposed:

$$V(\rho, \alpha) = V_1 + V_2 = \frac{1}{2}\rho^2 + \frac{1}{2}\alpha^2 \quad (5.61)$$

where ρ is the distance to the goal, and α the angle between the direction to the goal and the robot's longitudinal axis.

The time derivation of Eq. (5.61) over the paths defined by the set of kinematics equations is expressed as:

$$\dot{V}(\rho, \alpha) = \dot{V}_1 + \dot{V}_2 = -v_i cos(\alpha)\rho + \left(-\omega_i + v_i \frac{sin(\alpha)}{\rho} \right)\alpha \quad (5.62)$$

where v_i, ω_i are the desired translational and angular velocities, respectively.

For Eq. (5.62) to be negative definite, the following velocity functions are used:

$$v_i = k_v v_{\text{max}} \cos(\alpha) \tanh\left(\rho/k_\rho\right) \quad (5.63)$$

$$\omega_i = k_a \alpha + k_v v_{\text{max}} \frac{\text{th}\left(\rho/k_\rho\right)}{\rho} \cdot \frac{\sin(2\alpha)}{2} \quad (5.64)$$

where $\text{th}(\rho/k_p) \to 1$ if $\rho \to \infty$, k_p is the coefficient; $\frac{\text{th}(\rho/k_p)}{\rho} \to 1$ if $\rho \to 0$; k_α, k_v are positive coefficients.

The expressions in (5.63) and (5.64) can be used directly only if there are no obstacles and no limitations on the vehicles acceleration. In Ref. [39] the authors have proposed a modified version of the expression (5.61) based on the following objective function

$$G_m(v, \omega) = \lambda_1 (1 - |v - v_i|/(2v_{\max})) + \lambda_2 (1 - |\omega - \omega_i|/(2\omega_{\max}))$$
$$+ \lambda_3 \text{dist}(v, \omega) \qquad (5.65)$$

where $\lambda_i > 0$, $i = 1,2,3$ satisfying condition $\sum_i \lambda_i = 1$.

In Ref. [39] the authors obtained successful experimental results. Selection of the weighting coefficients in the function (5.64) and the procedure accounting for the sign change of the functions (5.62) and (5.63) is not obvious. The relationship between the obtained velocity value and the set of admissible velocities is not described explicitly.

In Ref. [40] the local minima problem is being solved using the Global Dynamic Window Approach. In this method the motion paths are planned using the information about the configuration of the areas free for the vehicle's motion. There is no need for a priori information about the environment. The planner obtains this data from the environment model or sensory information. In the process of motion, the obstacles location map is updated at each step of the planning task solution. The map is represented by a coordinate grid with its cells being assigned the values corresponding to the obstacles located in them.

The global dynamic window method presented in Ref. [40] combines the method of fast obstacle avoidance form Ref. [33] with the global navigation function $nf1$ from Ref. [41] that does not have local minima. In the classical form, this function is built using the wave algorithm generating wave propagation from the goal point.

The global dynamic window method assumes calculation of the function $nf1$ at each planning step, which enables this method to be used in unknown environments. The value of the function $nf1$ is calculated not for all the map cells but only for those that are located between the vehicle and the goal.

In the global dynamic window approach, the following multicriterial objective function is used

$$\Omega_g(p, v, a) = \alpha \cdot nf1(p, v) + \beta \cdot \text{vel}(v) + \gamma \cdot \text{goal}(p, v, a) + \delta \cdot \Delta nf1(p, v, a)$$
$$(5.66)$$

where p, v, α are the position, velocity and acceleration of the vehicle, respectively; α, β, γ, δ the positive coefficients; $nf1(p,v)$ the global navigational function having a maximum at the goal point; $\text{vel}(v)$ the trajectory velocity function; $\text{goal}(p,v,\alpha)$ the binary function equal to 1, if the path passes through the goal region, otherwise it is 0, $\Delta nf11$ indicates by how much the value of the $nf1$ will reduce during the next step.

The above-described methods are used in combination with the potential-field method [9,11,13] and the search methods like A*-algorithm for the full mapping of an a priori known environment or DF*-algorithm [42,43]—an improved version of A*-algorithm for mapping in unknown environments.

The idea behind D*-algorithm is that as the new information about the environment arrives, the built path is recalculated partially only in those sections where the changes are necessary, which increases numerical efficiency of the planning process.

Analogously to A*-algorithm, D*-algorithm uses open and closed lists. The states with higher and lower values of the path length function also get to the open list.

The path generation process ends when the open list receives the states with the path length function values greater or equal to the path length from the current vehicle's position.

FD*-algorithm differs from the classical D*-algorithm in using the function describing the path length from the checked cell to the vehicle cell. So the path length function changes only in those directions that mostly suit for the vehicle's motion.

An example of approach combining the FD*-algorithm with the dynamic window is presented in Ref. [44]. Selecting the best velocity value the authors use a measure from Ref. [45] describing how close a chosen path is to potential obstacles. This measure is considered only when the main safety condition for current path (distance to the nearest obstacle is more than length of braking path) is fulfilled, and it is described by the expression

$$
\vartheta_{\text{clear}}(v, \omega) = \begin{cases} 0, & npu \ t_{\text{col}} \leq T_b \\ \dfrac{t_{\text{col}} - T_b}{T_{\max} - T_b}, & npu \ T_b < t_{\text{col}} \leq T_{\max} \\ 1, & npu \ t_{\text{col}} > T_b \end{cases} \tag{5.67}
$$

where $T_b(v, \omega) = \max\left(\frac{v}{v_b}, \frac{\omega}{\omega_b}\right)$ is the braking time; $t_{\text{col}}(v, \omega) = \frac{\rho_{\min}(v,\omega)}{v}$ the time taken to collide with the nearest obstacle; $T_{\max} = \frac{s_i}{v_{\max}}$ is the maximal possible time to collide; $v_{\max}$ the maximal linear velocity; and s_i the sensor visibility range.

In Ref. [44] the following form of multicriterial objective function is used for the search for the optimal speed:

$$
\Gamma(v, \omega) = \lambda \vartheta_{\text{clear}} + (1 - \lambda) \vartheta_{\text{path}} \tag{5.68}
$$

where λ is the coefficient.

The measure of smoothness is given by the following expression [37]:

$$
\vartheta_{\text{path}}(v, \omega) = 1 - \frac{\sum_{i=1}^{Nt} \sum_{j=1}^{Np} j d_{ij} - D_{\min}}{D_{\max} - D_{\min}} \tag{5.69}
$$

where N_t is the number of points approximating the path curvature; N_p the number of points on the effective path; $d_{i,j}$ the distance between ith point of the path and jth point of the effective path; and $D_{\min} = \min_{(v,\omega)} \left\{ \sum_{i=1}^{Nt} \sum_{j=1}^{Np} j d_{ij} \right\}$;

$$D_{\max} = \max_{(v,\omega)} \left\{ \sum_{i=1}^{Nt} \sum_{j=1}^{Np} j d_{ij} \right\}.$$

In Ref. [44] the so-called effective path is introduced. Essentially, it is the straight line segment connecting the current robot position and the reference point on the path.

Maximum length of the effective path depends on $T_{\max}$ and is given by

$$R_{\max}(v_c) = (v_c + \dot{v}_a \Delta t) T_{\max} \tag{5.70}$$

where v_c is the current speed; $\dot{v}_a$ the maximal acceleration; and Δt the time interval.

Usually the reference point is located at the place of repeated change of the path direction with respect to the current vehicle's position. The minimum effective path length depends on the braking time $T_{b\;\max} = \frac{v_{\max}}{v_b}$

$$R_{\min}(v_c) = 0,5\dot{v}_b T_{b\;\max}^2. \tag{5.71}$$

Therefore, this approach improves the path shape achieved by the FD*-algorithm. However, selection of the number of points approximating the path and the effective path is performed empirically, which does not always lead to the optimal solution.

The Ref. [45] is an example of the combination of several approaches improving the planner. In this case three methods dynamic window approach, elastic band approach, and the global navigation function $nf1$ were combined together. The approach described in Ref. [45] provides smooth safe paths except for the cases of motion in narrow corridors close to other vehicles.

The algorithm presented in Ref. [24] belongs to a class of algorithms of randomized path planning. It differs from the one presented in Ref. [16] that uses randomly changed potential fields and probabilistic roadmap algorithm [21] by accounting for nonholonomic limitations imposed on the vehicle.

In Ref. [24] the path planning is performed as a search for a continuous path from an initial position x_{init} to a goal region $X_{goal} \subset X$ or goal state x_{goal}. It is assumed that a fixed obstacle region X_{obs} must be avoided when an explicit representation of X_{obs} is not available. States in X_{obs} correspond to vehicle's velocity bounds, configurations at which a robot is in collision with an obstacle, or other limitations, depending on the task.

A Rapidly-Exploring Random Tree (RRT) will be constructed so that all of its vertices are states in X_{free}, the complement of X_{obs}. Furthermore, all the path will lie entirely in X_{free}.

For a given initial state, x_{init}, an RRT, T, with K vertices is constructed as shown below:

GENERATE_RRT($x_{\text{init}}, K, \Delta t$)

1 T.init(x_{init});
2 for k =1 to K do
3     $x_{\text{rand}} \leftarrow$ RANDOM_STATE();
4     $x_{\text{near}} \leftarrow$ NEAREST_NEIGHTBOR(x_{rand},T);
5     $u \leftarrow$ SELECT INPUT(x_{rand}, x_{near});
6     $x_{\text{new}} \leftarrow$ NEW_STATE($x_{\text{near}}, u, \Delta t$);
7   T.add_vertex(x_{new});
8   T.add_edge(x_{near}, x_{new}, u);
9 Return T

Let ρ denote a distance metric on the state space. The first vertex of T is $x_{\text{init}} \in X_{\text{free}}$. In each iteration a random state, x_{rand}, is selected from the set X (it is assumed that X is bounded). Step 4 finds the closest vertex to x_{rand} in terms of ρ. Step 5 selects an input, u, that minimizes the distance from x_{near} to x_{rand} and ensures that the state remains in X_{free}. Collision detection can be performed by an incremental method such as Mirtich's V-clip [46]. NEW_STATE is called on each input to evaluate a potential new state (if U is not finite, it can be discretized, or an alternative optimization procedure can be used). The new state, x_{new}, which is obtained by applying u, is added as a vertex to T. An edge from x_{near} to x_{new} is also added, and the input u is recorded with the edge (because this input must be applied to reach x_{new} from x_{near}).

The key advantages of RRTs are (1) expansion of an RRT is heavily biased toward unexplored portions of the state space; (2) the distribution of vertices in an RRT approaches is predefined by a chosen function of random numbers generation (the sampling distribution) leading to a consistent behavior of the vehicle; (3) an RRT is probabilistically complete under very general conditions; (4) the RRT algorithm is relatively simple, which facilitates performance analysis; and (5) an RRT always remains connected even though the number of edges is minimal.

Considering the advantages of all algorithms described in Ref. [24], the questions still remain concerning the influence of Δt and ρ on the algorithm's effectiveness. It is also not clear how the performance will suffer from acting in complicated environments with non-convex obstacles. In general, probabilistic search algorithms do not guarantee generation of a path leading exactly to the goal. The final node of the tree is located at a certain allowed distance to the goal point. If the allowed value is small, the tree will have a large number of vertices and the search will consume a significant amount of time. Increasing the allowed distance to the goal increases the search speed by decreasing the number of vertices, which at the same time leads to lower accuracy.

In Ref. [47] an approach is proposed that corrects the initial path so that the final node of the tree matches the goal point. The idea of the approach is in introducing the perturbing functions added to control inputs of the system. The direction of perturbations is selected to reduce the Lyapunov function that includes distance from the tree node to the goal and the minimal distance to the obstacles at each point of the path. However, the method does not guarantee the local minima problem solution if the obstacle is located between the goal point and the final tree node.

In Ref. [48] a hierarchical approach to creation of a path planner is proposed. The path is generated incrementally for the unknown environments. The top level planner is built using A*-algorithm modified for working in unknown environments. The bottom layer planner models the vehicle's kinematics and accounts for the dynamic limitations to correct the commands of the top layer planner. The work proposes a situational approach to the vehicle motion planning where the search for the feasible and safe motions is performed in a reduced space of velocities.

5.3.3 Example of Planning Algorithm Implementation Accounting for the Vehicle's Inertia

In order to test the effectiveness of the planning algorithm accounting for the limitations put on the vehicle's motion a widespread approach is used—specifically, a combination of A*-algorithm and modified dynamic window approach. An optimal path is build using A*-algorithm in the explored section of the scene. The dynamic window method is used to modify the path so that the vehicle moving at a maximal possible velocity does not approach the obstacles.

The essence of the planning based on the A*-algorithm is as follows. An environment map is created. For this purpose the functioning plane is split into square cells. Each cell is assigned a value, marking it as free or occupied by an obstacle or being a starting (goal) cell. Initially all the cells are marked as free. As data starts to flow from the locator, the map is refreshed. A*-algorithm is started each time the change occurs in a cell located on a built path or when the motion parameters change.

The motion safety is ensured by marking all the cells adjacent to the obstacle as impassable. The number of impassable cells is determined by the cell size and the set minimal obstacle-vehicle distance. To avoid the problem of local minima, the cells visited several times are marked as occupied by obstacles.

After the map is created or updated, two lists of cells are created. The first one is called an OPEN list and it holds the cells requiring calculation of the vehicle path length. The second one is called a CLOSED list, holding the cells for which the path length is already calculated.

Then on the generated map, a cell with the vehicle is considered together with eight surrounding cells as shown in Fig. 5.23.

The starting cell where the next path section begins is added to the OPEN list. Then the cells surrounding it are checked. The cells are added to the OPEN list if they do not have obstacles in them and they do not cross the boundaries of the scene. For each of the cells added to the OPEN list, the coordinates of the previous (parent) cell are stored and the function f is calculated defining the path

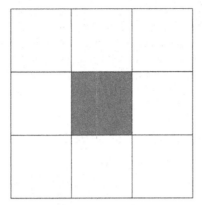

FIGURE 5.23 Block of cells ready to join the OPEN list.

length through the current cell. The starting cell is added to the CLOSED list. Then a cell with a lowest value of f is taken out of the OPEN list and so on.

The value of f function is calculated as follows:

$$f = g + h \qquad (5.72)$$

where g is the cost of the path from the starting cell to this one, and h the cost of the path from the current cell to the goal.

The value of g is determined as a sum of vehicle's cost of motion from the starting cell to the parent one and from the parent cell to the current position. For example, for the case of square cells as shown in Fig. 5.23, we assume that from the central cell the cost of motion of a cell shifting either vertically or horizontally is equal to 10 and the cost of motion of a cell shifting diagonally is equal to 14.

The value h is determined as a cost of horizontal and vertical motion between the current cell and the goal. It is equal to 10 times the number of passed cells.

A*-algorithm of path planning can be presented as follows:
1. Add the starting cell to the OPEN list.
2. Repeat the following steps:
   a. Search the OPEN list for cell with the lowest value of f, name it a current cell;
   b. Put the current cell into the CLOSED list and delete it from the OPEN list;
   c. For all the eight surrounding cells.
      i. if there is an obstacle in the cell or it is in the CLOSED list, the cell is ignored;
      ii. if the cell is not in the OPEN list, it is added to it; the current cell becomes a parent cell for the considered cell; the values of f, g, and h are calculated for the considered cell.

iii. if the cell is already in the OPEN list, the length of a path through this cell is checked; the values of g is used for this purpose; if the value of g is reduced, this means that the path will become shorter; in this case the current cell becomes a parent for the considered cell and the values of g and f are recalculated;

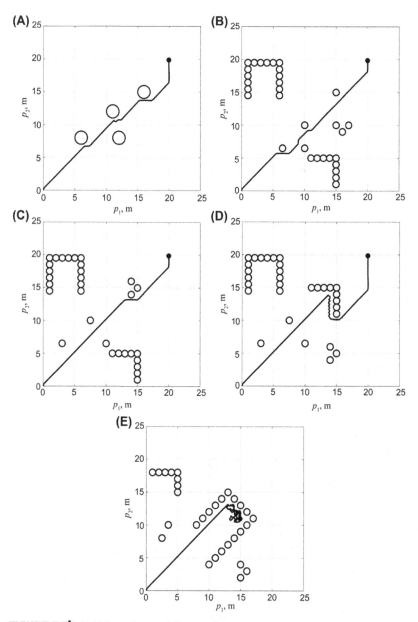

FIGURE 5.24 Vehicle motion modeling results.

    d. Stop if the goal cell is added to the OPEN list (in this case the path is planned) or the OPEN list is empty and the goal point is not reached (the path is impossible to build).

3. Save the path.

The limitations are accounted for by selection of the maximal speed out of the set obtained by calculation of the expression (5.59).

The modeling results for the vehicle's motion in the preset of five scenes accounting for the dynamic limitations are presented in Fig. 5.24.

The obtained experiment results are collected in Table 5.4. The method's effectiveness is estimated using the four indexes introduced in Chapter 1.

The modeling results show the effectiveness of the vehicle path planning approach based on A*-algorithm and the dynamic window method. The proposed hybrid approach allows the vehicle to reach the goal even in complicated scenes excluding the ones with U-shaped obstacles.

Table 5.4 Experimental Values of Performance Indexes

Criterion	a_n	A*-Algorithm Based Planner
Safety criterion S_m	0.2	0.83
		0.79
		0.78
		0.88
		0
Path length P_L	0.2	32.24
		31.66
		30.33
		39.95
		∞
Task completion time t_m	0.2	31.65
		31
		30
		37.8
		∞
Mission success coefficient F	0.4	1
		1
		1
		1
		0

5.4 SUMMARY

Several widespread motion planning methods were considered for obstructed environments in this chapter. In particular, we considered methods using potential fields, Voronoi diagrams, and search algorithms accounting for vehicle's inertia. To certain extent, each of the methods uses the graph theory and the analytical expressions in description of the vehicle's motion. The main difference between the methods considered in this chapter and earlier is that the planning task for unknown environments is solved by graph-analytical methods without any artificial intelligence technologies.

The effectiveness of the considered methods is estimated in implementation of the planner in a junction with a position-path controller. At each modeling step, the planners form the desired requirements to the paths in the space of working coordinates in a form of linear and quadratic forms (Eq. 1.6).

An analysis of existing planners based on potential-field method has shown that this method is mostly used for path planning in environments with stationary obstacles. This is connected to the fact that effectiveness of the specific potential function essentially depends on the configurations and types of obstacles. It should be mentioned that it seems promising to set the potential functions, so that they include the vehicle's and obstacle's safety radiuses because this reduces computational costs of path planning. Using the normalized potential functions reduces the task completion time since this optimizes the vehicle's motion speed in unknown environments. The potential trigonometric functions have shown the maximal effectiveness in the five modeled scenes but this does not guarantee the positive result under different conditions.

The planner based on Voronoi diagrams has a low computational effectiveness which can be partially compensated by mapping. However, the positive results obtained by other researchers call for the future researches in this domain directed toward increasing its computational effectiveness.

The survey of planners accounting for the vehicle's inertia has shown that dynamic window and velocity curves approaches are the most promising since they ensure the path smoothness and satisfy all kinematic and dynamic requirements put on the vehicle. Most frequently, these methods are used in combination with the search algorithms. A distinctive feature of the considered methods is that in each planning stage they form not a single path but a set of paths consisting of multitude of allowed motion paths. This allows the planner to solve the task even in environments with dynamic obstacles. The modeling results have shown that the length of the vehicle's motion path essentially depends on the size of the coordinate grid cell, and motion with a maximal possible speed is achieved.

For all the considered graph-analytical methods, it should be mentioned that the planner's functionally and effectiveness can be significantly increased by developing combined methods and allowing to apply them to the environments with non-stationary obstacles.

REFERENCES

[1] A.K. Platonov, A.A. Kirilchenko, M.A. Kolganov, Potentials' Method in the Task of Path Selection: History and Possibilities, Institute of Applied Mathematics Named After M.V. Keldysh, Moscow, 2001.

[2] D.E. Koditschek, Task encoding: toward a scientific paradigm for robot planning and control, Robotics and Automation Systems 9 (1992) 5–39.

[3] E. Rimon, D.E. Koditschek, The construction of analytic diffeomorphisms for star worlds, IEEE International Conference on Robotics and Automation Proceedings 1 (1989) 21–26. Waschington etc.

[4] K. Fisk, D. Keski, L. West, Printed circuit boards automated design, Institute of Electrical and Electronics Engineers Writings 55 (1967) 217–228.

[5] A.K. Platonov, I.I. Karpov, A.A. Kirilchenko, Potentials' Method in the Routing Task, Institute of Applied Mathematics AN SSSR, Moscow, 1974.

[6] Y. Ichikawa, M. Fujie, N. Ozaki, On mobility and autonomous properties of mobile robots, Robot (1984) 31–36.

[7] R.A. Brooks, A.M. Flynn, T. Marill, Self Calibration of Motion and Stereo Vision for Mobile Robot Navigation, Fourth International Symposium of Robotics Research, MIT Press, Santa Cruz, CA, 1987, pp. 267–276.

[8] O. Khatib, Real-time obstacles avoidance for manipulators and mobile robots, International Journal of Robotics Research 5 (1) (1986) 90–98.

[9] C. Pozna, R.E. Precup, L.T. Koczy, A. Ballagi, Potential field-based approach for obstacle avoidance trajectories, The IPSI BgD Transactions on Internet Research 8 (2) (2002) 40–45.

[10] A. Ferrara, M. Rubagotti, Sliding mode control of a mobile robot for dynamic obstacle avoidance based on a time-varying harmonic potential field, in: ICRA Workshop: Planning, Perception and Navigation for Intelligent Vehicles, 2007.

[11] M.A. Padilla Castaneda, J. Savage, A. Hernandez, F. Arambula Cosío, Local autonomous robot navigation using potential fields, in: X.-J. Jing (Ed.), Motion Planning, InTech, 2008, ISBN 978-953-7619-01-5.

[12] S.S. Ge, Y.J. Cui, New potential functions for mobile robot path planning, IEEE Transactions on Robotics and Automation 16 (5) (2000) 615–620.

[13] Y. Koren, J. Borenstein, Potential field methods and their inherent limitations for mobile robot navigation, International Conference on Robotics and Automation 2 (1991) 1398–1404.

[14] T. Uusitalo, S.J. Johansson, A reactive multi-agent approach to car driving using artificial potential fields, in: Proceedings of IEEE Conference on Computational Intelligence and Games, 2011, pp. 203–210.

[15] J.F. Canny, M.C. Lin, An opportunistic global path planner, in: Proceedings of IEEE International Conference on Robotics and Automation, 1990, pp. 1554–1559.

[16] S. Shimoda, Y. Kuroda, K. Iagnemma, High speed navigation of unmanned ground vehicles on uneven terrain using potential fields, Robotica 25 (4) (2007) 409–424.

[17] F. Li, Y. Tan, Y. Wang, G. Ge, Mobile Robots path planning based on evolutionary artificial potential fields approach, in: Proceedings of the 2nd International Conference on Computer Science and Electronics Engineering, 2013, pp. 1314–1317.

[18] L. Tang, S. Dian, G. Gu, et al., A novel potential field method for obstacle avoidance and path planning of mobile robot, in: Proceedings of the 3rd International Conference on Computer Science and Information Technology (ICCSIT), vol. 10, 2010, pp. 633–637.

[19] P. Bhattacharya, M.L. Gavrilova, Roadmap-based path planning using the Voronoi diagram for a clearance-based shortest path, IEEE Robotics & Automation Magazine (2008) 58–66.

[20] H. Nolborio, T. Naniwa, S. Arimoto, A quadtree-based path-planning algorithm for a mobile robot, Journal of Robotics Systems 4 (4) (1990) 555–574.

[21] N. Amato, Y.A. Wu, A randomized roadmap method for path and manipulation planning, in: Proceedings of IEEE International Conference on Robotics and Automation, 1996, pp. 113–120.
[22] K. Kedem, M. Sharir, An efficient motion planning algorithm for a convex rigid polygonal object in 2-dimensional polygonal space, Discrete & Computational Geometry 5 (1) (1990) 43–75.
[23] J. Kim, R.A. Pearce, N.M. Amato, Extracting optimal paths from roadmaps for motion planning, in: Proceedings of IEEE International Conference on Robotics and Automation, 2003.
[24] S.M. La Valle, Rapidly-exploring Random Trees, a New Tool for Path Planning, 1998.
[25] E. Masehian, M.R. Amin-Naseri, A Voronoi diagram-visibility graph-potential field compound algorithm for robot path planning, Journal of Robotics Systems 21 (6) (2004) 275–300.
[26] C. Kimberling, Central points and central lines in the plane of a triangle, Mathematics Magazine 67 (1994) 163–187.
[27] J. Kim, F. Zhang, M. Egerstedt, A provably complete exploration strategy by constructing Voronoi diagrams, Autonomous Robots 29 (3) (2010) 367–380.
[28] S. Garrido, L. Moreno, D. Blanco, Exploration of 2D and 3D environments using Voronoi transform and fast marching method, Journal of Intelligent and Robotic Systems 55 (2009) 55–80.
[29] M. Seda, V. Pich, Robot motion planning using generalized Voronoi diagrams, in: Proceedings of the 8th WSEAS International Conference on Signal Processing, Computational Geometry and Artificial Vision, Greece, 2008, pp. 215–220.
[30] K. Weiler, P. Atherton, Hidden surface removal using polygon area sorting, in: Proceedings of the 4th Annual Conference on Computer Graphics and Interactive Techniques, 1977, pp. 214–222.
[31] M. de Berg, M. van Kreveld, M. Overmars, O. Schwarzkopf, Computational Geometry: Algorithms and Applications, Springer, 2000.
[32] B. Siciliano, O. Khatib, Springer Handbook of Robotics, Springer-Verlag, Berlin, 2008, pp. 827–851.
[33] D. Fox, W. Burgard, S. Thrun, The dynamic window approach to collision avoidance, IEEE Robotics and Automation Magazine 4 (1) (1997) 23–33.
[34] J. Minguez, L. Montano, Abstracting any vehicle shape and the kinematics and dynamic constraints from reactive collision avoidance methods, Autonomous Robotics 20 (1) (2006) 43–59.
[35] J. Minguez, L. Montano, Robot navigation in very complex dense and cluttered indoor/outdoor environments, in: Proceedings of the 15th IFAC World Congress, 2002.
[36] A. Bemporad, A. De Luca, G. Oriolo, Local incremental planning for car-like robot navigating among obstacles, in: Proceedings of the IEEE International Conference on Robotics and Automation, 1996, pp. 1205–1211.
[37] R. Simmons, The curvature-velocity method for local obstacle avoidance, in: Proceedings of IEEE International Conference on Robotics and Automation, 1996.
[38] R. Simmons, The lane-curvature method for local obstacle avoidance, in: IEEE RSJ International Conference on Intelligent Robots and Systems, 1998.
[39] H. Berti, A.D. Sappa, O.E. Agamennoni, Improved dynamic window approach by using Lyapunov stability criteria, Latin American Applied Research 38 (2008) 289–298.
[40] O. Brock, O. Khatib, High-speed navigation using the global dynamic window approach, in: Proceedings of IEEE International Conference on Robotics and Automation, 1999, pp. 341–346.
[41] J. Barraquand, J.C. Latombe, Robot motion planning: a distributed representation approach, International Journal of Robotics Research 10 (6) (1991) 628–649.
[42] K. Macek, I. Petrovic, E. Ivanjko, An approach to motion planning of indoor mobile robots, in: Proceedings of the IEEE International Conference on Industrial Technology, 2003, pp. 969–973.
[43] A. Stentz, The focused D* algorithm for real-time replanning, in: Proceedings of the International Joint Conference on Artificial Intelligence, 1995.

[44] M. Seder, K. Macek, I. Petrovic, An integrated approach to real-time mobile robot control in partially known indoor environments, in: Proceedings of the 31st Annual Conference of IEEE Industrial Electronics Society, 2005.

[45] R. Philippsen, R. Siegwart, Smooth and efficient obstacle avoidance for a tour guide robot, in: ICRA'03. IEEE International Conference on Robotics and Automation, 2003.

[46] B. Mirtich, V-clip: fast and robust polyhedral collision detection, ACM Transactions on Graphics 17 (3) (1998) 177–208.

[47] F. Lamiraux, E. Ferre, E. Valee, Kinodynamic motion planning: connecting exploration trees using trajectory optimization methods, in: Proceedings of 2004 IEEE Intern. Conf. on Robotics and Automation, vol. 4, 2004, pp. 3987–3992.

[48] X.C. Lai, S.S. Ge, A.A. Mamun, Hierarchical incremental path planning and situation-dependent optimized dynamic motion planning considering accelerations, IEEE Transactions on Systems, Man, and Cybernetics, Part B, Cybernetics 37 (6) (2007) 1541–1554.

Chapter Six

Motion Planning and Control Using Bionic Approaches Based on Unstable Modes

V. Pshikhopov, M. Medvedev

Southern Federal University, Taganrog, Russia

6.1 NON-FORMALIZED ENVIRONMENTS WITH POINT OBSTACLES

Vehicles continue finding their way into various fields of human activity and the requirements of their control systems algorithms become stronger. This is especially true for vehicles moving in non-formalized environments with stationary and non-stationary obstacles. From our point of view, the main problems of synthesis and implementation of such control systems are caused by a number of factors [1]. The first one is the practice to split the multiply connected control systems. On one hand, this simplifies the synthesis procedure and the structure of the control system. On the other hand, it holds us from obtaining the required performance of the closed-loop system. For example, usage of PID controller in the longitudinal and transverse control channels of a flying vehicle is justified only in the presence of a pilot in the control contour compensating the dynamic effects not accounted in the synthesis procedure. Application of the classical approaches to vehicle's controller synthesis can lead not only to a failure to reach the necessary quantitative performance requirement put on the path-tracking quality, but also can lead to a loss of system's qualitative qualities such as stability. Obviously, violation of assumptions taken in controller synthesis for the simplified model results in a closed-loop system losing the required accuracy, and if there are limitations put on the control actions, this can lead to instability.

The second factor giving the possibility to organize the necessary character of the vehicle's motion in the obstructed environment is the absence of effective methods of coupling the motion planning subsystems (strategic level of control

Path Planning for Vehicles Operating in Uncertain 2D Environments. http://dx.doi.org/10.1016/B978-0-12-812305-8.00006-5

system) with the tactical control level. The known approaches to vehicle's control systems design use approximation units for solution and interpolation of inverse kinematics problems. This introduces additional errors in the calculation of reference values for actuators and, therefore, into tracking of the planned trajectories. Besides, incorporation of additional units into the structure of control system can lead to lowering the reliability of the closed-loop control system. Various structural-algorithmic implemental aspects of coupling in the motion planning subsystem based on neural-like structures and neural networking system with position-trajectory controllers are considered in Refs. [2—5].

The presented synthesis results are based on the notions of structural synthesis theory by L.M. Boychuk published in 1960s—1970s particularly in Ref. [6] and developed the presented approaches for the class of mobile robots. In Ref. [7] it is proposed to transform nonstationary and stationary obstacles into repellers embracing them by attracting manifolds—attractors. In this way it becomes possible to perform transition from one stable state to another stable state using unstable modes.

In this chapter, the procedure of repellers organization is considered in the sense of definition given in Ref. [7].

Assume that the vehicle is described by two first equations of the system (Eq. 1.1). Then, all the multitude of the requirements to the steady state motion mode in the space $R^{n \times n}$ of the basic coordinates y and speeds $\dot{y}$ can generally be presented in the form of vector functions $\psi_{tr, sp}$ of the basic coordinates and orientation angles, and their derivatives of the form (1.6) and (1.7).

The control task is stated as follows. For a vehicle described by the two first equations of the system (Eq. 1.1), it is necessary to synthesize a control law $F_u(\delta)$ that will ensure asymptotic stability manifolds of Eqs. (1.6) and (1.7) transforming them into attractors. The obstacle Π with its characteristic point described by the vector $P_P = |p_{1p} \quad p_{2p} \quad p_{3p}|^T$ is transformed into a repeller, i.e., a repelling manifold, with a bypass radius not exceeding the value of r.

It should be mentioned that unlike the task statement formulated in Ref. [7], to calculate the obstacle avoidance trajectory in this case the values of speed and acceleration of obstacle's motion are not required.

The results used were also obtained in Refs. [7,8]. In Ref. [8] it is shown that if we select the Lyapunov vector function in the form

$$L = \left(\Psi^T \Psi + \dot{\Psi}^T D \dot{\Psi} \right)^2, \tag{6.1}$$

where $\Psi = \Psi_{tr} + A\Psi_{sp}$, and if the condition $TA = D$ is satisfied, the derivative of the function (6.1) is equal to

$$\frac{dL}{dt} = -2(T+A)\dot{\Psi}^T \dot{\Psi} \left(\Psi^T \Psi + \dot{\Psi}^T TA\dot{\Psi} \right). \tag{6.2}$$

From the expression (6.2) it follows that the vehicle's stable motion along the trajectory Ψ is ensured if the matrices T and A are positively definite.

It gives rise to the following question. Based on the stated task, how to transform an obstacle Π, described by the vector of coordinates y_Π of its characteristic point P_p, into a repeller?

This problem can be solved if in a certain zone r for the point P_p, one or two matrices T and A should be negatively definite. Without loss of generality, assume that $T = A = \mathrm{diag}\, S_i$, $i = \overline{1, \nu - 1}$, where S_i is a certain functional parameter given by the following expression:

$$S_i = a_{oi} - \frac{r_i^2}{\varepsilon^2} \tag{6.3}$$

where $\varepsilon^2 = (y_1 - y_{1p})^2 + (y_2 - y_{2p})^2$.

From the expression (6.3) it follows that for the final values of the constants a_{oi} and R_i in the zone of large deviations from the obstacle, the value of parameter S_i approaches the value of the constant a_{oi} that determines the dynamics of the vehicle's motion with respect to the manifold Ψ, i.e., $\lim_{\varepsilon \to \infty} S_i = a_{oi}$. In the obstacle zone, the expression $\lim_{\varepsilon \to 0} S_i = -\infty$ holds, which transforms the obstacle Π into an unstable point i.e., a repeller. The stability boundary is determined by the following obvious expression

$$r_i = \frac{r_i}{\sqrt{a_{oi}}} \tag{6.4}$$

Thus, forming the elements of matrices T and A according to the relation of the form (6.3), we ensure the stable character of the vehicle's motion along the manifold Ψ excluding the r vicinity of the point P_p.

For $\nu = 3$ the vehicle motion trajectory is determined by the intersection of two quadratic forms. The speed of attraction to them can be set individually selecting the parameters a_{oi}.

Let us consider an example of avoiding a stationary point obstacle by a vehicle based on a wheeled cart.

It is necessary to perform a vehicle's motion along a straight line with a trajectory speed of $V_k = 3$ m/s. The obstacle's position is known.

The modeling results for the autonomous vehicle control system are presented in Figs. 6.1 and 6.2.

The procedures of repellers organization in the state space of the vehicle presented in Ref. [7] allow us to perform correct motion of the vehicle in a priori, non-formalized environment with point obstacles. Unlike the traditional approaches, the obtained structural-algorithmic solutions don't require knowing the values of speeds and accelerations of the obstacles. The obtained results can be used in the group control tasks, in the control systems of unmanned flying vehicles, etc.

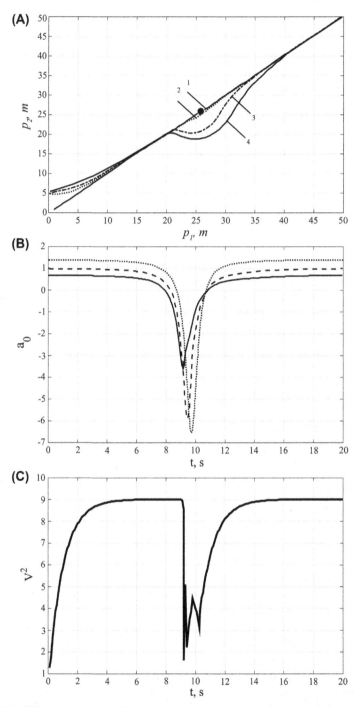

FIGURE 6.1 Modeling results. (A, B) Mobile robot's motion trajectory and the value of co-efficient s for $r = 4$, $y_{1\Pi} = 27$, $y_{1\Pi} = 28$; 1, desired trajectory; 2, $a_0 = 1.4$; 3, $a_0 = 1.0$; 4, $a_0 = 0.7$. (C, D) Value of the squared trajectory speed V^2 for $a_0 = 1.4$ and $a_0 = 0.7$. (E, F) Value of control u_1 for $a_0 = 1.4$ and $a_0 = 0.7$.

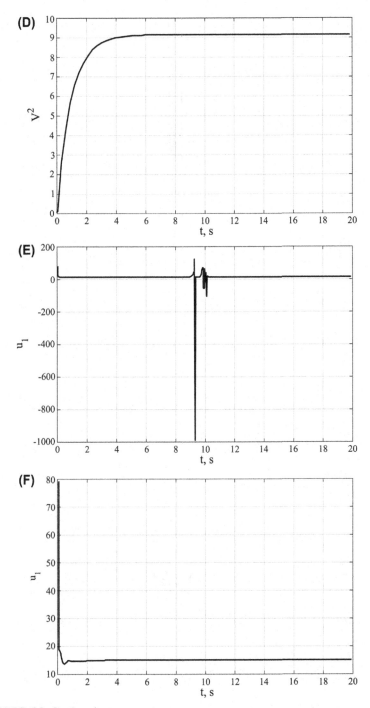

FIGURE 6.1 Continued.

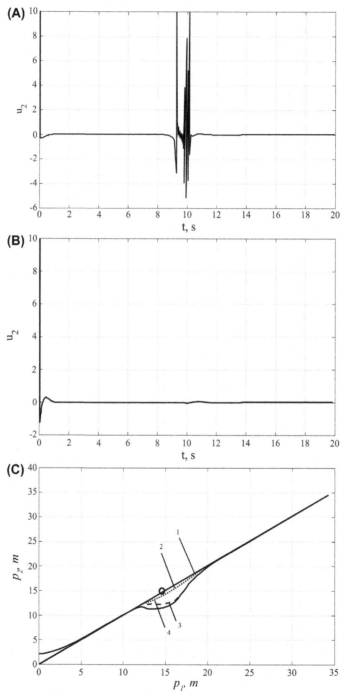

FIGURE 6.2 Modeling results (continuation). (A, B) Control values u_2 for $a_0 = 1.4$ and $a_0 = 0.7$. (C, D) Vehicle's motion trajectory and values of the coefficient s for $a_0 = 2$, $x_P = 15$, $y_p = 16$; 1 - desired trajectory; $2 - r = 5$; $3 - r = \sqrt{30}$; $4 - r = \sqrt{35}$.

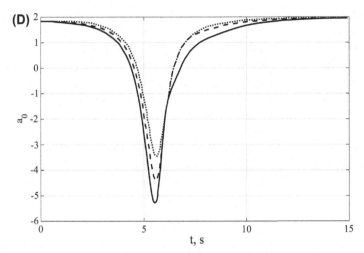

FIGURE 6.2 Continued.

6.2 NON-FORMALIZED ENVIRONMENTS WITH COMPLICATED OBSTACLES

6.2.1 Vehicle's Trajectories Generation

The problem of vehicle's motion organization in a priori, non-formalized environments is addressed in a series of works of Russian [9−11] and other [12−14] authors. The proposed solutions are implemented in a variety of intelligent systems of motion planning (strategic level) and motion control as well as in the methods of tactical control algorithms creation. For example, in the works based on the potential-field method [14] the proposed control algorithms assume preliminary planning of the vehicle's motion paths (which is not always possible in an unpredictable environment [10]). The control law consists of two components: one stabilizes the motion trajectory and the other one deforms it in the vicinity of an obstacle. In Ref. [12] the concept laid in the foundation of the certainty grid method requires mapping of the vehicle's local work area. And this sets additional requirements for the sensor system. The algorithms of control generation presented in Ref. [15] assume preliminary mapping of work area and determining the dimensions and motion parameter of the obstacles.

Thus, we can identify the main problem of the mentioned approaches and path planning methods. It is the necessity for trajectory preplanning or mapping of the vehicles work area. This poses essential limitations on the motion speed in a priori, non-formalized environments and causes high computational expenditures.

In this section, based on the results obtained in Ref. [1,3,7], we discuss an approach to control synthesis for a vehicle functioning in a priori, non-formalized environments. In the process, the Lyapunov instability theorem [16,17] is used, which lets us loosen the requirements at the strategic level and the vehicle's sensory system.

More generally, based on Refs. [1,3,7], stationary and mobile obstacles Π_j can be located in the vehicle's work area. The vehicle's (y_1,y_2) position in the reference coordinate system is known. The goal point coordinates (y_{f1},y_{f2}) are set in the same coordinate system. Generally, the point A_f is mobile. It is assumed that the vehicle's sensor system can find the shortest distance r_c, $c = \overline{1,k}$ from the vehicle's characteristic point (e.g., center of the fixed coordinate system) to the nearest points belonging to one or several obstacles Π_i located within the sensor's range.

For the vehicle described by the first two equations of the system (1.1) it is necessary to synthesize control F_u ensuring vehicle's motion from an arbitrary point (y_{01},y_{02}) to the set goal point (y_{f1},y_{f2}) satisfying the following conditions

$$r_c \geq r \qquad (6.5)$$

where r_c is the distance to the nearest obstacle and r is a constant, setting allowed shortest distance from the vehicle's characteristic point to any of the obstacles Π_j.

Unlike the task set in Ref. [1,7], in this case there is no need to obtain the values for the obstacle's speed and acceleration. There are no limitations put on the number of obstacles of their motion pattern and there is no need to define any characteristic points coordinates. So, the obstacle is represented by a geometrically complicated shape but its geometrical characteristics are not used for avoidance, and the technical vision system gives out only the distance r_c.

6.2.2 Control System Structural and Algorithmic Implementation

The essence of the proposed approach to structural and algorithmic implementation of the vehicle's control system is in using control actions stabilizing trajectories (Eqs. 1.6 and 1.7) in the areas free form obstacles, and application of the third Lyapunov theorem (theorem on instability) when violating inequality (Eq. 6.5), i.e., when the vehicle is located at the distance r_c, less than the allowed value r.

The vehicle can be either in the area free from the obstacles satisfying the inequalities (Eq. 6.5) or in the area where they are violated. It is proposed to organize the vehicle's motion modes so that the planned trajectories set by the manifolds of Eqs. (1.6) and (1.7) would be stable in the first case and unstable in the second.

In the study by Boychuk [18] and Zimin [19], it is shown that inequalities (Eq. 6.5) can be represented by one equality and here it is proposed to form it as follows:

$$\beta = \sum_j |r_c - r| + \sum_j (r_c - r) \qquad (6.6)$$

where j is the number of nearest points within the range of the sensor system belonging to one or several obstacles.

Obviously, if all the inequalities (Eq. 6.5) are satisfied, the value of parameter β (Eq. 6.6) will be equal to zero and will take only positive nonzero values if any of the inequalities (Eq. 6.5) is violated.

Another study by the author [7] shows that stable pattern of the vehicle's motion along ψ_{tr} trajectory is provided by positive definiteness of T and A matrices. Consequently, if conditions (Eq. 6.6) are broken, one or both T and A matrices shall be negatively definite. Without loss of generality we can assume that $T = A = \text{diag } s_i$, $i = \overline{1, \nu} T = A = \text{diag} s_i$, $i = \overline{1, \nu}$, where s_i is a functional parameter, setting the definiteness property for T and A matrices.

In view of the above, we propose to set T and A matrices elements in the form of the following function

$$
s_i = \begin{cases} s_0 = \text{const} > 0, & \text{for} \quad \beta = 0, \\ -\dfrac{1}{\beta}, & \text{for} \quad \beta \neq 0, \end{cases} \tag{6.7}
$$

where s_0 sets the motion character in the area free from the obstacles.

Forming T and A matrices elements according to the relation in the form of (6.7), we will provide stable character of vehicle's motion along the Ψ_{tr} manifold except for the areas where inequalities (Eq. 6.5) are violated.

Since it is planned to organize the vehicle's motion from one stable state through unstable motion into another stable state, β is called as bifurcation parameter according to Refs. [7,20].

The structural scheme of the closed-loop system, realizing the proposed control algorithm, is given in Fig. 6.3. It corresponds to the structure given in Fig. 1.11 and includes the following elements: units M, F_d, $\Sigma(\Theta, x)$, adder, integrators, and links, reflecting the plant structure (Eq. 1.1); vehicle motion planner (operator, onboard computer, neural network, etc.) designed for forming manifold's coefficients in Eqs. (1.6) and (1.7) and $\tilde{T}$ and $\tilde{A}$ matrices elements; sensors of the internal D_x and external D_y coordinates; unit of control

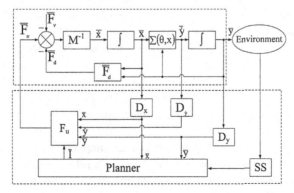

FIGURE 6.3 Closed-loop system structure.

actions calculation F_U; technical vision system (TVS), determining the distance r to the nearest points, located in the vehicle's sensor system range and belonging to one or several obstacles. The closed-loop system functions in the following way. Upon control system initialization, TVS unit defines r_c distance, calculates the value of bifurcation parameter β based on (Eq. 6.6), and then transmits it to the planner to form T and A matrices elements.

Planner forms motion parameters vector I; goal point coordinates A_f, generally a time dependent one; matrices elements A_i, defining requirements for the set vehicle's motions (at the initial moment in the case of obstacles absence A_i A_i matrices elements set the straight line connecting initial vehicle's position to point A_f); desired velocity of the object motion V^*; s_i parameters, defined according to (Eq. 6.7) and setting vehicle's motion pattern in the transient modes; ξ parameter characterizing control task form; and, if needed, evaluation vectors $\widehat{l}$ and $\widehat{F_v}$ design parameters and external disturbances [21].

Unit F_u, based on I parameters received from the planner, forms control actions F_U, which are transmitted to the vehicle's actuating devices and provide vehicle's motion along the straight line connecting vehicle's initial position and the point A_f.

If conditions (Eq. 6.5) are violated, the planner changes values of s_i parameters according to the fornula (6.7), and the vehicle changes to unstable motion mode, till the bifurcation parameter β sets to zero, i.e., until the object enters area free of obstacles.

Upon entering the area free of obstacles, the planner forms matrices elements A_i and the object turns till the condition $\alpha_A - \alpha < \frac{\pi}{2}$ is fulfilled, where α_A, α are the angles of direction toward the goal point A_f and current heading, respectively. If bifurcation parameter β equals to zero, the planner forms matrices elements A_i corresponding to the straight line connecting current vehicle position with A_f point, and F_u unit provides vehicle's motion along the newly planned straight line. In another case at $\beta \neq 0$, the object changes to the unstable motion mode till bifurcation parameter β sets to zero.

It should be noted that organization of vehicle's turn is used for additional definition of the motion direction along the newly planned straight line. In the case of using other procedures of forming requirements for trajectory velocity V [9], the turn stage can be excluded from the proposed algorithm.

When the vehicle enters the neighborhood of the goal point A_f free from the obstacles, the planner reforms matrices elements A_i.

From the described vehicle control system functioning algorithm it follows that there are limitations on its usage, for example, in case of intended blocking of object motion by the obstacles or other vehicles the plant can stay in unstable motion mode; when the obstacles have a significantly complicated shape, such as a labyrinth, then the task set for the object cannot be solved within the limits of the proposed algorithm which is shown further in modeling results.

The main advantages of this approach are simplicity of its realization and that during obstacles avoidance organization in a priori, non-formalized environment there is no need in plotting the trajectories in obstacles area which, in many cases related to dynamic changes in environment, is not always possible in real time.

6.2.3 Modeling Results

Control system modeling is performed by a terrestrial robot as given in Fig. 1.2 and its kinematic model described by the first equation of the system (Eq. 1.2). Position-path control law is used for controller Eqs. (1.9), (1.10), and (1.13) with conditions of Eqs. (1.18) and (1.19).

Modeling results of position-path control system using bionic approach based on unstable modes for scene 1 are given in Figs. 6.4 and 6.5. Fig. 6.4 gives the robot motion trajectory with obstacle sensor area highlighted. Obstacles on the scene are represented by the circles and their visible part, detected by the sensor is marked by a bold line.

Fig. 6.5 gives T and A matrices' parameters, which take negative values when approaching the obstacles. Let us note that the control system switches into unstable mode four times. Relay switching of bifurcation parameter is typical for unstable modes and this is obviously connected to the vehicle's motion along the trajectory defined by the boundary of inequality (Eq. 6.5).

From the modeling results it follows that the vehicle has successfully completed the mission. For the set task, the values of the performance criteria, described in Chapter 1, are given in Table 6.1.

Analogous modeling results for the scene 2 are given in Figs. 6.6 and 6.7. Zero initial conditions were selected. Sensor model is analogous to the one from scene 1. Fig. 6.6 gives vehicle's motion trajectory, and Fig. 6.7 gives change of T and A matrices parameters.

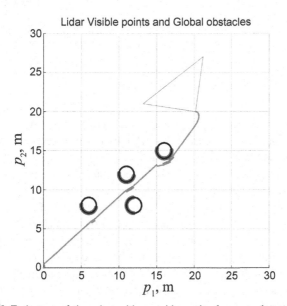

FIGURE 6.4 Trajectory of the robot with unstable modes for scene 1.

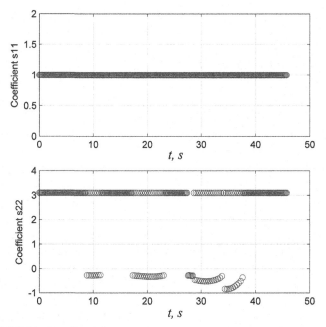

FIGURE 6.5 Matrices T and A parameters.

Table 6.1 Experimental Values of Motion Criteria for Bionic Methods Based on Unstable Modes

Comparison Criteria	a_n	Scene 1	Scene 2	Scene 3	Scene 4	Scene 5
Safety criterion S_m	0.2	0.36665	0.38771	0.82494	0.0058909	0.10215
Path length P_L	0.2	44.7932	32.9213	35.623	47.9744	Inf
Task completion time t_m	0.2	45.72	34.38	36.72	49.86	Inf
Mission success coefficient F	0.4	1	1	1	1	0

The modeling results show that scene 2 is passed successfully by the robot, using bionic approaches based on unstable modes in its control system. As in Fig. 6.4, the visible parts of the obstacles are highlighted by a bold line.

The task completion criteria, as in the previous case, are given in the second column of Table 6.1.

Figs. 6.8 and 6.9 give robot's trajectory and change of T and A matrices' parameters with respect to time and robot orientation angle change during motion for scene 3. The robot has also completed the task successfully. Herewith, the control system switches to unstable mode only once, when passing the last group of obstacles (Fig. 6.8). All modeling parameters correspond to the parameters for scenes 1 and 2. Performance criteria values are given in the third column of Table 6.1.

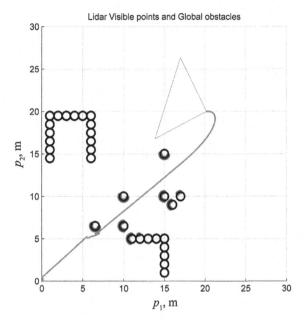

FIGURE 6.6 Robot trajectory with unstable modes for scene 2.

Figs. 6.10 and 6.11 give robot trajectory and change of T and A matrices parameters with respect to time for scene 4. The robot has also completed the task successfully. All modeling parameters correspond to the parameters for the previous scenes 1, 2, and 3. Performance criteria values are given in the fourth column of Table 6.1.

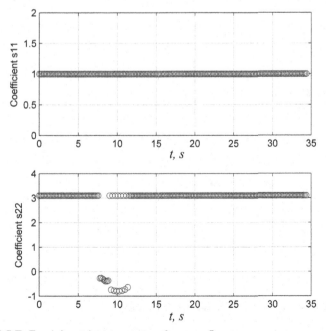

FIGURE 6.7 T and A matrices parameters for scene 2.

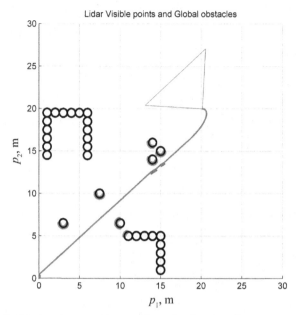

FIGURE 6.8 Robot trajectory with unstable modes for scene 3.

Figs. 6.12 and 6.13 give robot's path and the change of T and A matrices parameters with respect to time for the scene 5. All modeling conditions match the ones for the scenes 1, 2, 3, and 4.

As opposed to the previous scenes, robot does not complete the task in this scene. Performance criteria values are given in the fifth column of Table 6.1. This is caused by the location of the obstacles in the scene creating a local

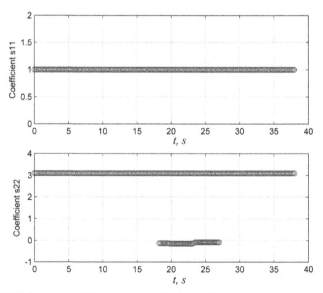

FIGURE 6.9 T and A matrices parameters for scene 3.

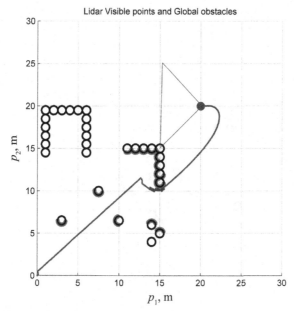

FIGURE 6.10 Robot trajectory with unstable modes for scene 4.

minimum. The situation requires application intelligent control technologies for the robot to get out.

Thus, the presented modeling results confirm the efficiency of bionic approach to the motion path planning.

Main advantages of the given method are as follows:

- low requirements to the sensor system, it is enough to know only the distance to the obstacle at the current point of time;
- minor computational efforts, resulting from the absence of mapping, image processing, etc.; and

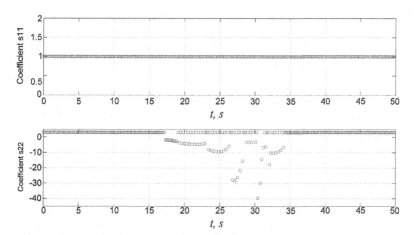

FIGURE 6.11 T and A matrices parameters for scene 4.

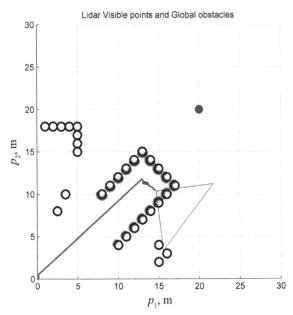

FIGURE 6.12 Robot trajectory with unstable modes for scene 5.

- relatively simple integration into any existing closed-loop control systems circuits, involving correcting of robot reference motion model parameters.

 Limitations of the suggested approach include the fact that in presence of certain obstacles (narrow dead-ends, bottlenecks, etc.) the method can fall

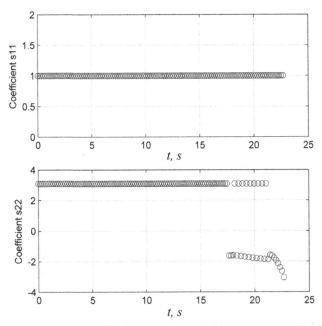

FIGURE 6.13 *T* and *A* matrices parameters for scene 5.

into the local minimum of the error function. This disadvantage can be eliminated by introducing an additional decision-making contour.

Note that the method suggested in this section allows avoidance of mobile obstacles also. For example, Figs. 6.14−6.16 show the results of modeling of mobile obstacle avoidance having radius equal to 1 m, starting its motion from the point having coordinates (5, 5) at the initial moment. Obstacle motion

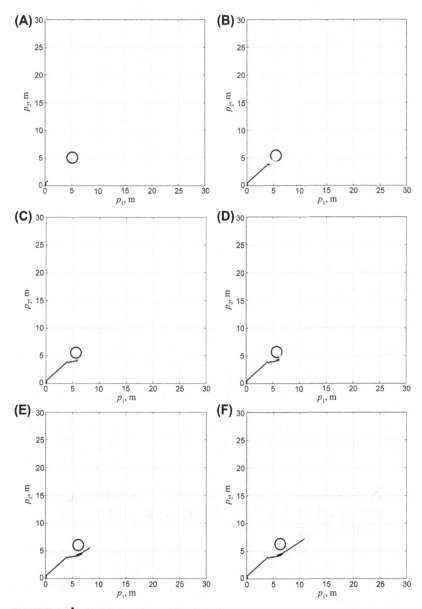

FIGURE 6.14 Avoidance of a mobile obstacle.

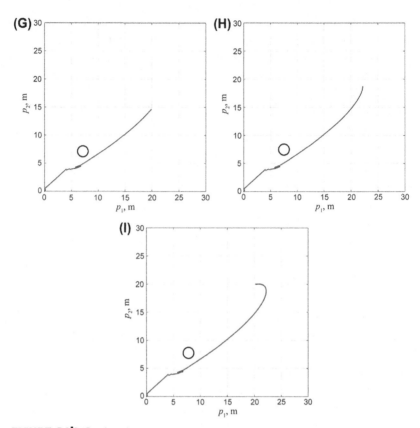

FIGURE 6.14 Continued.

velocity is 0.1 m/s and the trajectory is linear, at an angle of 45° to P_1 axis. Obstacle avoidance process is shown in Fig. 6.14. Resulting trajectory with illustration of initial and final positions of obstacle is shown in Fig. 6.15. Changing of controller parameters is given in Fig. 6.16.

Analogous modeling results for the obstacle traveling at 0.3 m/s are given in Fig. 6.17–6.19.

From the given modeling results it can be seen that in spite of substantial obstacle velocity (up to 30% of the maximum vehicle motion), method of unstable modes allows successful functioning in a nonstationary environment.

6.3 COORDINATED APPLICATION OF UNSTABLE MODES AND VIRTUAL POINT FOR OBSTACLE AVOIDANCE

Ensuring effective functioning of mobile robots is a rather urgent problem while organizing their motion in an undetermined environment. Depending on the character of the task being solved, robot features, and environment, function criteria can vary.

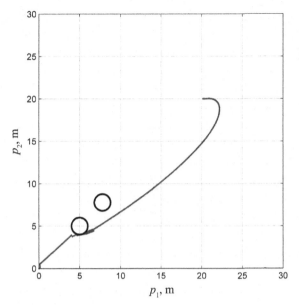

FIGURE 6.15 Resulting path.

The vehicle control method for undetermined environments proposed in Ref. [1,7] allows decreasing requirements for planning subsystem and sensor support. However, as shown in Section 5.2.3, there are possible situations of such obstacles location, when the suggested method does not provide deadlocks recovery (see Fig. 6.12).

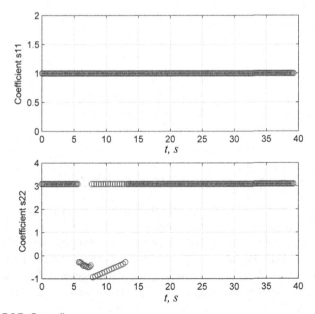

FIGURE 6.16 Controller parameters.

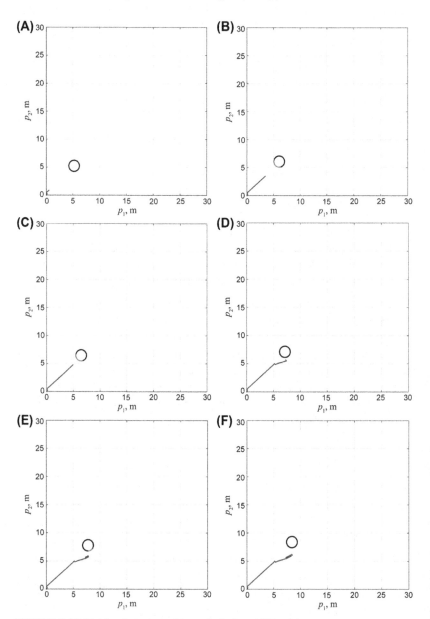

FIGURE 6.17 Mobile obstacle avoidance (velocity of 0.3 m/s).

It should be noted that the given disadvantages are characteristic of many planning methods whether using mapping or not. So, for example, studies [22,23] and Chapter 5 of this book represent artificial potential field for local planning methods. Potential-field method being one of the most widely used path planning methods also has some disadvantages caused by local minima problem and oscillations in areas crowded with obstacles [23] as shown in Figs. 5.9 and 5.10.

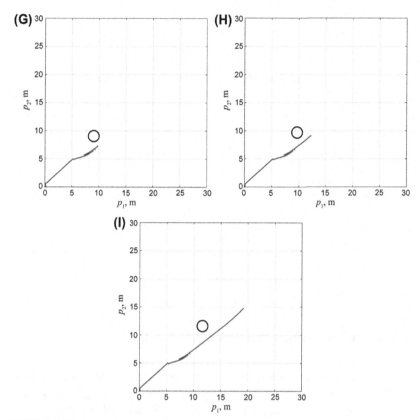

(G)

(H)

(I)

FIGURE 6.17 Continued.

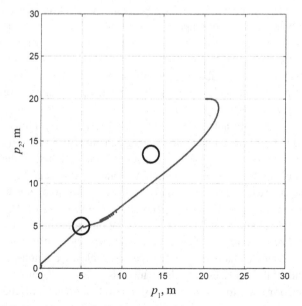

FIGURE 6.18 Resulting path (velocity of 0.3 m/s).

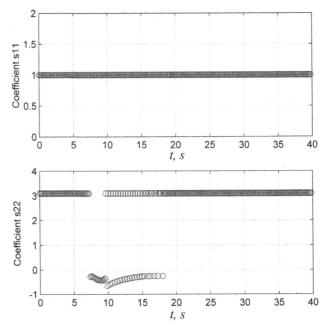

FIGURE 6.19 Controller parameters (velocity of 0.3 m/s).

As a rule, these disadvantages are eliminated by means of specifying stringent requirements for sensor subsystems and onboard computers, provided by the necessity of online mapping of the robot work area. Most frequently, the reasons of the indicated disadvantages are obstacles of Π-type, as it follows from modeling results presented in the previous chapters of this book.

In this section, the mentioned disadvantage is eliminated by algorithms which combine path planning and it's following. Basic controller is generated using position-path control method. The unstable mode is used by introducing the bifurcation parameter in Eq. (6.7).

In addition to the unstable modes, obstacles avoidance involves obstacles and virtual point aggregation procedures. Obstacles aggregation procedure is described by an algorithm given in Fig. 6.20.

The goal and obstacle coordinates are received upon control system initialization, the obstacle images are aggregated, and the bifurcation parameter calculation unit determines the distances r_c and calculates the value of bifurcation parameter β. Then it is fed to the planner that generates the parameters of the matrices T and A. If the conditions of Eq. (6.5) are violated, the planner changes the values of the parameters s_i according to the expression (6.7), and the vehicle enters the unstable motion mode until the bifurcation parameter β zeroes and the vehicles enters the area clear of obstacles.

After the vehicle enters the area with no obstacles, the planner recalculates the elements of the matrices A_i in the manifolds of Eqs. (1.6) and (1.7), and the vehicle performs a turn as a result of control actions until the condition

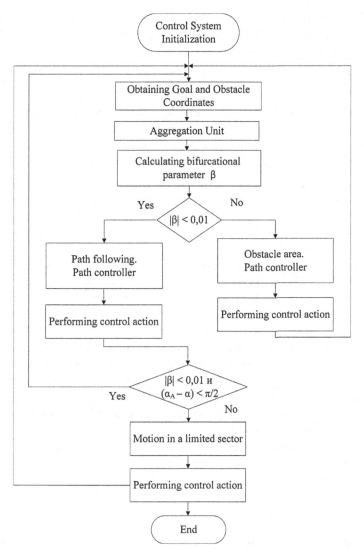

FIGURE 6.20 Algorithm of planner operation with obstacles aggregation procedure.

$\alpha_A - \alpha < \pi/2$ is satisfied, where α_A, α are the angle of directions to the goal and the current heading, respectively. If bifurcation parameter β equals to zero the planner forms matrices element A_i corresponding to the straight line connecting the current vehicle's position with A_f, and F_u unit provides vehicle's motion along the newly planned straight line.

Otherwise with $\beta \neq 0$, the vehicle changes to unstable motion mode till bifurcation parameter β sets to zero.

The suggested algorithm does not eliminate all the disadvantages related to vehicle's falling into local minima. In this regard, we proposed robot's motion control using the concept of a virtual goal point [24–26] formed from initial goal point coordinates by means of vector $[p_{1f}, p_{2f}]$, turn by the angle γ. Angle γ value is a function of bifurcation parameter β, which depends on the distance between the robot and the goal:

$$\gamma = k_\gamma \beta. \tag{6.8}$$

Virtual goal point coordinates are defined by the following formula:

$$\begin{bmatrix} p_{1f\_v} \\ p_{2f\_v} \end{bmatrix} = \begin{bmatrix} \cos(\gamma) & -\sin(\gamma) \\ \sin(\gamma) & \cos(\gamma) \end{bmatrix} \begin{bmatrix} p_{1f} \\ p_{2f} \end{bmatrix} \tag{6.9}$$

where p_{1f_v}, p_{2f_v} are the coordinates of virtual goal point; k_γ is adjustable parameter.

When $|\beta| > 0.01$ condition is violated, the vehicle changes to the virtual goal point mode.

Fig. 6.21 represents the goal point changing into virtual one using formulas (6.8) and (6.9). The manifold of requirements to the steady state mode of the vehicle's motion in the $R^n \times {}^n$ environment of basic coordinates y and $\dot{Y}$ velocities is represented in the form of a vector function ψ for basic coordinates and orientation angles and their derivatives.

Using the concept of virtual goal point given in Fig. 6.22, the vehicle control system structure has a strategy level, a planner, which develops the desired trajectories, based on sensor information and task type; a tactical level, controller, which calculates the necessary controls based on the desired set trajectory; and a unit of virtual goal point, which develops virtual points from goal point coordinates by means of a turn.

The planner using the concept of a virtual goal point is described by the following formula:

$$\Psi_{tr} = \begin{cases} \left| \begin{matrix} P + A_3(t) \\ \phi - \phi^* \end{matrix} \right| = 0, & \text{if } \beta = 0, \\[4ex] \left| \begin{matrix} P + \begin{bmatrix} \cos(\gamma) & -\sin(\gamma) \\ \sin(\gamma) & \cos(\gamma) \end{bmatrix} A_3(t) \\ \phi - \phi^* \end{matrix} \right| = 0, & \text{if } \beta > 0, \end{cases} \tag{6.10}$$

where P is a vector of the robot external coordinates and $A_3(t)$, matrix of coefficients, defining desirable trajectory of the robot's motion.

Selection of controller type and adjustment of its parameters in accordance to the motion goal and sensor information are performed by the planner, the operation algorithm for which is explained in Fig. 6.23.

Upon control system initialization, technical vision system defines r_c distance and calculates the value of bifurcation parameter β according to formula (6.6) and then transmits it to the planner to form T and A matrices elements.

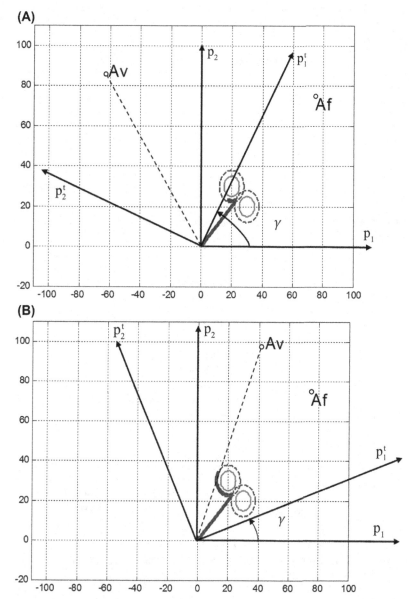

FIGURE 6.21 Change of goal point into virtual point.

The planner also forms the following motion parameters: goal point A_f coordinates, A matrices elements, defining requirements for the set vehicle's motions, desired velocity of the object motion V^*. The controller forms control actions F_u, which are transmitted to the vehicle actuating devices and provide vehicle motion along the straight line connecting vehicle's initial position and A_f point.

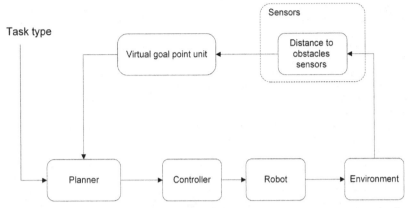

FIGURE 6.22 Flowchart of a position-path control system using the virtual goal point concept.

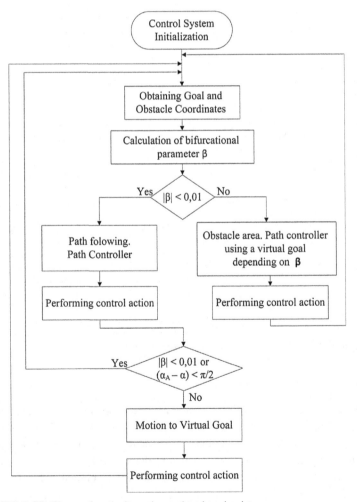

FIGURE 6.23 Planner functioning using a virtual goal point.

Algorithm studies have shown that in the case of the obstacle having narrow "bottlenecks" a situation may arise when the robot enters into an infinite loop and cannot reach the goal point. This problem can be solved by conversion matrix multiplication by K vector, and the conversion is given by the following formula:

$$\begin{bmatrix} p_{1f\_v} \\ p_{2f\_v} \end{bmatrix} = \begin{bmatrix} k_1 \\ k_2 \end{bmatrix} \begin{bmatrix} \cos(\gamma) & -\sin(\gamma) \\ \sin(\gamma) & \cos(\gamma) \end{bmatrix} \begin{bmatrix} p_{1f} \\ p_{2f} \end{bmatrix}, \tag{6.11}$$

where p_{1f_v}, p_{2f_v} virtual goal point coordinates; k_1, k_2, constants; $k_1, k_2 > 1$.

Advantages of the suggested method:

- when organizing obstacles avoidance in a priori, non-formalized environment there is no need to form paths in the obstacles' neighborhood, which in some cases, based on environment dynamic changes, is not always possible in real time;
- possibility to organize robot motion in a non-determinate environment with closely located convex obstacles;
- possibility of motion in the environment with obstacles of L-, U-, and E-shapes;
- requirements for trajectory planning systems are reduced; and
- limitations for shape, velocity, motion, and number of obstacles are decreased.

Let us study the usage of unstable modes method using virtual point for modeling the scenes presented in Fig. 1.3.

Modeling results for scene 1 are given in Figs. 6.24–6.26. As seen from the given results, the robot enters obstacle avoidance mode only once, repulsing

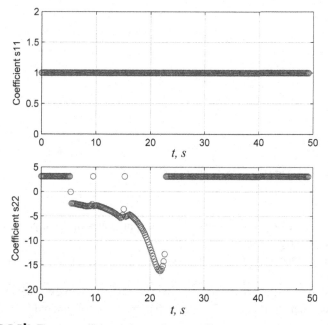

FIGURE 6.24 The *ss* coefficients change for scene 1.

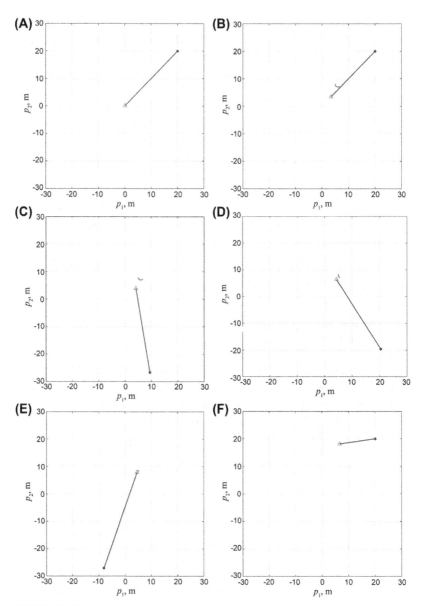

FIGURE 6.25 Virtual point position change for scene 1.

toward the left side of the scene, and then approaches the goal point along the scene upper edge.

Performance criteria for the scene 1 are given in the first column of Table 6.2.

Modeling results for scene 2 are given in Figs. 6.27–6.29. Negative values of s22 coefficient in Fig. 6.27 indicate that the robot has entered into obstacles

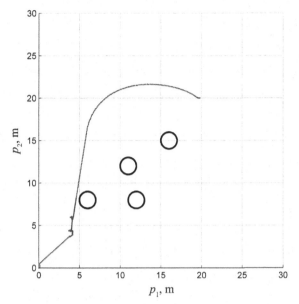

FIGURE 6.26 Robot's motion path for scene 1.

avoidance mode several times. Resulting path with the complete obstacles configuration on the map is given in Fig. 6.28. Virtual point turn is shown in Fig. 6.29.

The performance criteria of robot motion for scene 2 are given in the second column of Table 6.2.

Modeling results for scene 3 are given in Figs. 6.30–6.32. As Fig. 6.30 shows, while being in operation the robot has entered into obstacle avoiding mode several times. Goal point turn is shown in Fig. 6.32. Robot resulting trajectory is given in Fig. 6.31.

First obstacle avoidance lasts from the 2nd to 16th second and corresponds to detection of the first obstacle having coordinates (3; 6.5) m. Second period starts at 19th second and ends at the 35th second of operation. During this

Table 6.2 Experimental Values of Motion Criteria for Bionic Methods With Unstable Modes With a Virtual Point

Comparison Criteria	a_n	Scene 1	Scene 2	Scene 3	Scene 4	Scene 5
Safety criterion S_m	0.2	0.5752	1.713	0.5001	0.7087	0.6947
Path length P_L	0.2	37.308	36.0448	45.9943	38.4093	37.7628
Task completion time t_m	0.2	49.14	53.46	68.76	67.68	64.98
Mission success coefficient F	0.4	1	1	1	1	1

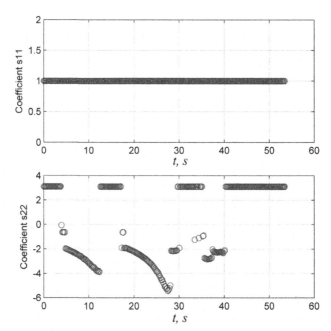

FIGURE 6.27 The *ss* coefficients change for scene 2.

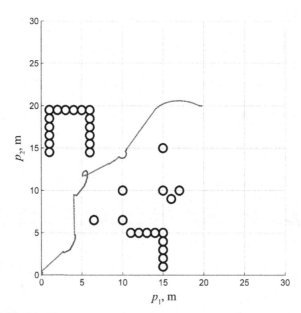

FIGURE 6.28 Resulting trajectory of robot motion for scene 2.

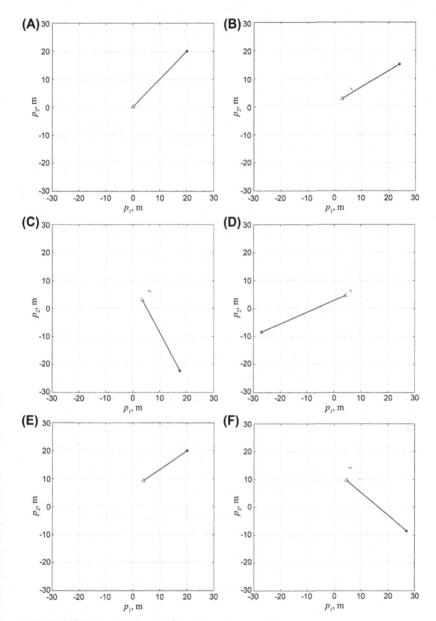

FIGURE 6.29 Virtual point position change for scene 2.

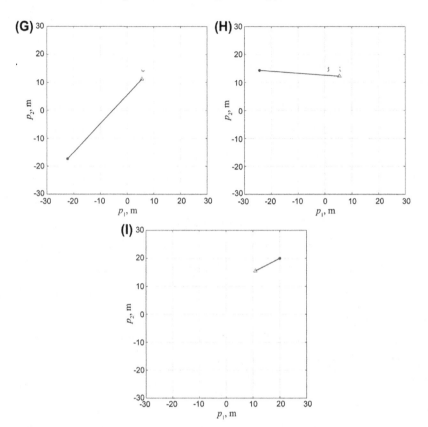

FIGURE 6.29 Continued.

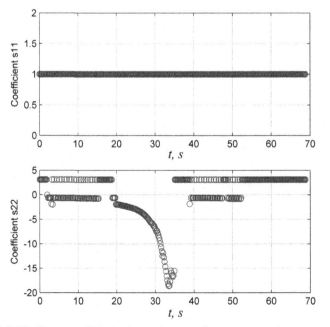

FIGURE 6.30 The *ss* coefficients change for scene 3.

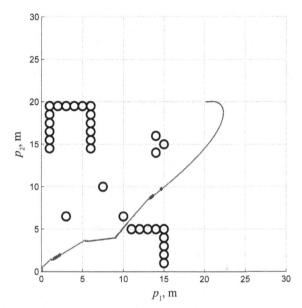

FIGURE 6.31 Resulting trajectory of robot motion for scene 3.

period the robot passes between the edge of angular and isolated obstacles. The final period corresponds to the avoidance of the group of three obstacles in the upper right corner of the scene.

The performance criteria of robot's motion for scene 3 are given in the third column of Table 6.2.

The modeling results for scene 4 are given in Figs. 6.33—6.35. As seen from Fig. 6.33, negative coefficients, setting obstacle avoidance mode with formation of a virtual point, appear at the 19th second of robot motion. The robot switches to the basic controller operation mode upon passing obstacles zone at the 54th second. Change of virtual point position in time is shown in Fig. 6.33. Resulting trajectory on the scene with indication of the obstacles points, defined by laser scanner, is shown in Fig. 6.34.

Performance criteria of robot motion for scene 4 are given in the fourth column of Table 6.2.

Modeling results for scene 5 are shown in Figs. 6.35—6.37. Graphs for the system speed coefficients values are shown in Fig. 6.35, Fig. 6.37 illustrates the virtual point action with an indication of the moment of time counting from the start of the robot's motion. The robot position (orientation is not shown) is marked with a triangular marker. Large dot marker indicates the virtual point position, and a line connecting robot position and virtual point is shown for convenience.

As seen from Figs. 6.35—6.37, the vehicle enters into unstable mode of obstacles avoidance with a turn of virtual point at approaching to Π-shaped

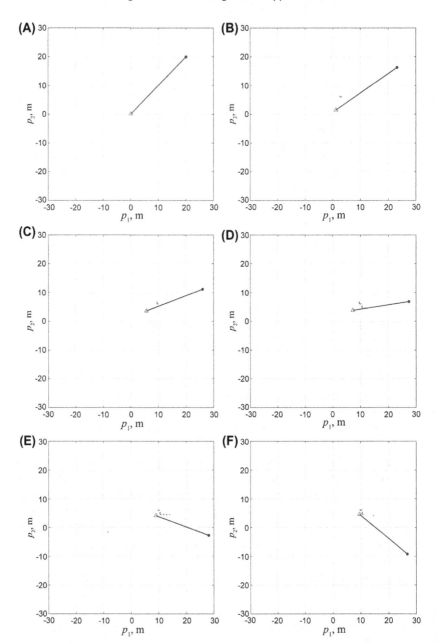

FIGURE 6.32 Virtual point position change for scene 3.

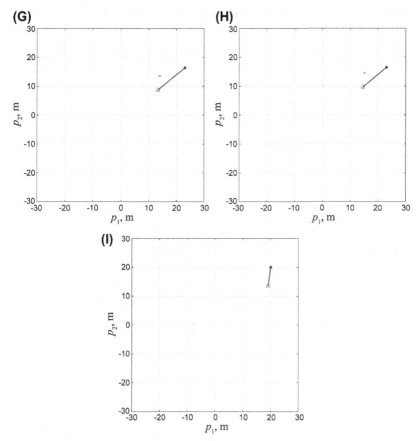

FIGURE 6.32 Continued.

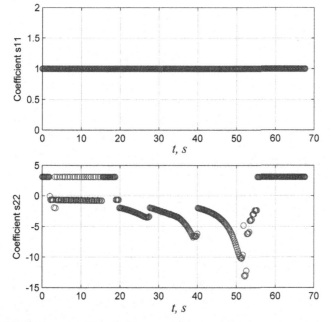

FIGURE 6.33 The *ss* coefficients for scene 4.

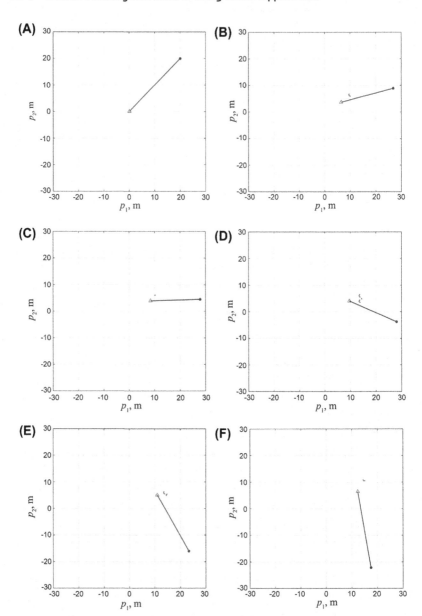

FIGURE 6.34 Virtual point position change for scene 4.

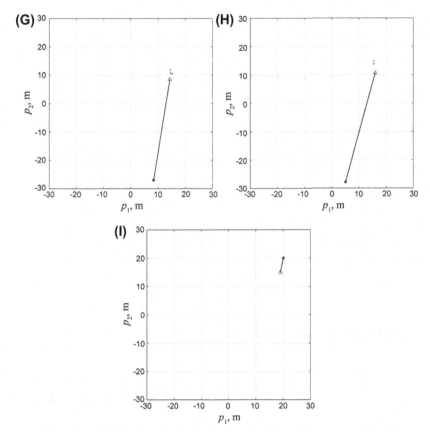

FIGURE 6.34 Continued.

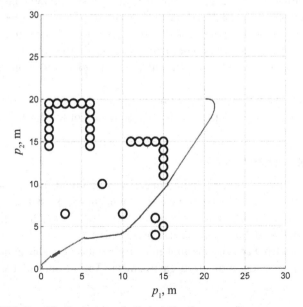

FIGURE 6.35 Resulting trajectory of robot motion for scene 4.

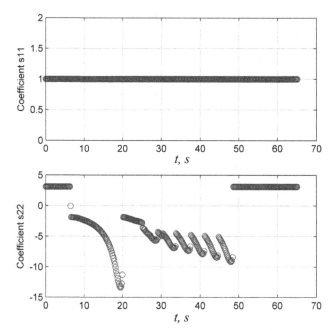

FIGURE 6.36 The ss coefficients change for scene 5.

obstacle approximately at the 7th second of the motion. The robot comes out of the obstacle avoidance mode at the 49th second and moves toward the goal point.

Performance criteria of robot motion for scene 5 are given in the fifth column of Table 6.2.

Comparing the results of position-path control system modeling based on unstable modes and unstable modes using virtual point, it can be noted that introduction of virtual point allowed solving problem of falling into local minimums. And as it can be seen from the modeling results, those unstable modes in control system with virtual point are observed more frequently (Fig. 6.38).

Comparative analysis of the obtained modeling results with other vehicle path planning methods is presented in the summary of this book.

6.4 SUMMARY

This chapter gives structural-algorithmic control solutions for vehicles moving in obstructed environment. In vehicle motion planning, the stationary and mobile obstacles are converted into repellers by means of synthesized controls. Synthesis and modeling examples are presented.

The considered approach is applied to organization of vehicle motion planning in a priori, non-formalized environment. The novelty of the proposed

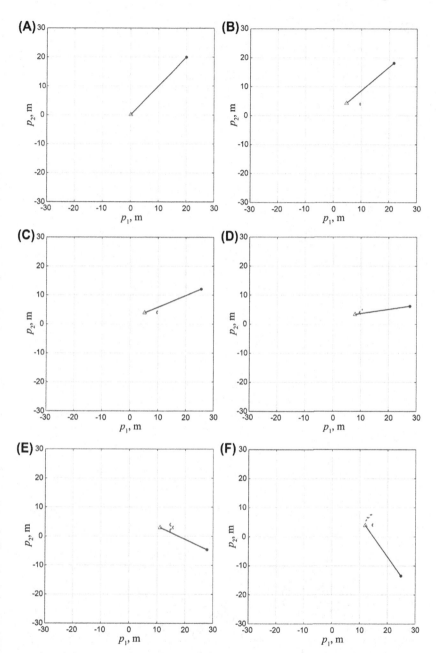

FIGURE 6.37 Virtual point position change for scene 5.

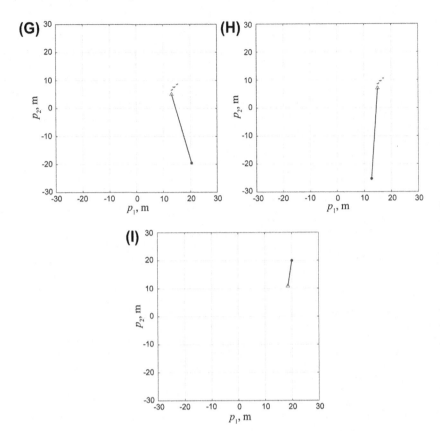

FIGURE 6.37 Continued.

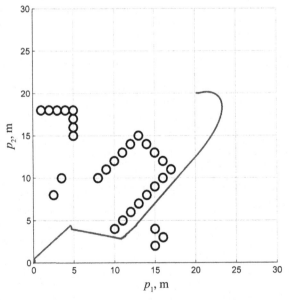

FIGURE 6.38 Resulting trajectory of robot motion for scene 5.

solutions is in the introduction of a bifurcation parameter in order to form an unstable motion mode while moving from one stable state to another one. The proposed approach reduces the requirements of the planning and control subsystems. It does not require mapping and also eliminates the need for a complicated navigation system. The chapter presents modeling results for a wheeled cart robot functioning in the environments with stationary and mobile goal points and obstacles of various forms.

REFERENCES

[1] V. Kh Pshikhopov, Attractors and repellers in design of mobile object control systems, Proceedings of TSURE 3 (2006) 117–123. Special issue "Perspective systems and control problems", Taganrog.

[2] V. Kh Pshikhopov, M. Yu Sirotenko, Structural-algorithmic realization of control system of the autonomous mobile robot with neural network planner, Proceedings of TSURE 3 (2004) 185–191. Special issue "Intelligent CADs", Taganrog.

[3] Yu V. Chernuhin, V. Kh Pshikhopov, S.N. Pisarenko, O.E. Trubachev, Software environment for modeling of the actions of adaptive mobile robots having two-level control system, Mechatronics 6 (2000) 26–30.

[4] Yu V. Chernuhin, V. Kh Pshikhopov, S.N. Pisarenko, O.E. Trubachev, Hardware and software system for modeling of intelligent mobile robots neural network control systems, Mechatronics 1 (2002) 27–29.

[5] V. Kh Pshikhopov, Yu V. Chernukhin, M. Yu Medvedev, Structural synthesis of dynamic regulators for position-path adaptive mobile robots control systems on base of mini-airships, in: Proceedings of VII International SAUM Conference on Systems, Automatic Control and Measurements, Vrnjachka Banja, Yugoslavia, 2001, pp. 64–69.

[6] L.M. Boychuk, Structural Synthesis of Multi-layer Automated Functional Systems of Dynamic Objects Control, Preprint 74-23, Kiev, Cybernetics Institute AN USSR, 1974.

[7] V. Kh Pshikhopov, Repellers organization during mobile robot motion in the environment with obstacles, Mechatronics, Automation and Control 2 (2008) 34–41.

[8] V.Kh Pshikhopov, M.Yu Medvedev, Mobile Object Control in Definite and Indefinite Environments, Nauka, Moscow, 2011.

[9] S.F. Burdakov, I.V. Miroshnik, R.E. Stelmakov, Wheeled Robots Motion Control Systems, Science, Saint Petersburg, 2001.

[10] I.M. Makarov, V.M. Lokhin, Intelligent Systems of Automated Control, Fizmatlit, Moscow, 2001.

[11] P.D. Krutko, P.A. Osipov, Kinematic algorithms of the mobile robots transport systems motion control, Proceedings of the Russian Academy of Sciences, Control Theory and Systems 3 (1999) 153–160.

[12] J. Borenstein, Y. Koren, Real-time Obstacle Avoidance for Fast Mobile.

[13] C. Canudas de Wit, C.H. Khennouf, C. Samson, O.J. Sordalen, Nonlinear Control Design for Mobile Robots, Recent Trends in Mobile Robots, World Scientific, 1993, pp. 121–156.

[14] O. Khatib, Real-time obstacles avoidance for manipulators and mobile robots, International Journal of Robotics Research 5 (1) (1986) 90–98.

[15] B.V. Topchiev, Mobile robots synergetic control, Non-linear World 2 (4) (2004) 239–249.

[16] B.P. Demidovitch, Lectures on Mathematical Theory of Stability, Nauka, Moscow, 1967.

[17] A.M. Lyapunov, General Task on Motion Stability, Cherepovets: Merkurij-press, 2000.

[18] L.M. Boychuk, Method of Structural Synthesis of Automated Control Systems, Energy, Moscow, 1971.

[19] M.F. Zimin, Concerning the equations determining squares, volumes and their boundaries, Mathematic Education 1 (1930).

[20] V. Kh Pshikhopov, Mobile object control in a priory non-formalized environments, Izvestiya SFedU. Engineering Sciences, Taganrog 12 (2008) 6−9.

[21] V. Kh Pshikhopov, M. Yu Medvedev, Structural Synthesis of Mobile Object Autopilot Systems with Disturbances Estimation, 1, Information-Measuring and Control Systems, Moscow, 2006, pp. 103−109.

[22] J. Borenstein, Y. Koren, Real-time obstacle avoidance for fast mobile robots in cluttered enviroments, in: IEEE International Conference on Robotics and Automation, 1990, pp. 572−577.

[23] Y. Koren, J. Borenstein, Potential field methods and their inherent limitations for mobile robot navigation, in: International Conference on Robotics and Automation, vol. 2, 1991, pp. 1398−1404.

[24] A.S. Ali, R.V. Fedorenko, V.A. Krukhmalev, Autonomous mobile robot skif-3 control system for nonformalized environment, Izvestiya SFedU. Engineering Sciences 3 (2010) 132−143.

[25] V. Kh Pshikhopov, A.S. Ali, Avoidance of local minima of error function during robot motion in indefinite environment, Proceedings of the Universities of the North-Caucasian Region, Technical Sciences 6 (2011) 26−31.

[26] V. Kh Pshikhopov, A.S. Ali, Avoidance of Local Minima of Error Function Based on Position-path Control During Robot Motion in Indefinite Environment, All-Russian School of Sciences "Microelectronic Information-Control Systems and Complexes", Novocherkassk, 2011, pp. 175−179.

Summary

Finally, let us carry out a comparative analysis of the following trajectory planning methods considered in this book:
- Potential-field methods (criteria values as in Table 5.2);
- Voronoi diagram method (criteria values as in Table 5.3);
- reactive navigation method (criteria values as in Table 5.4);
- control method using fuzzy systems (criteria values as in Table 3.5);
- Distance vector histogram (DVH) method in a traditional basis (criteria values as in Table 2.2);
- DVH method in a formal neuron basis (criteria values as in Table 2.3);
- neural network hybrid method (criteria values as in Table 2.4);
- genetic search without mapping (criteria values as in Table 4.8);
- genetic search with mapping (criteria values as in Table 4.14);
- planning method using unstable modes (criteria values as in Table 6.1); and
- combined planning method using unstable modes and virtual goal point (criteria values as in Table 6.2).

To normalize modeling results, let us average the performance criteria for all the scenes. The reciprocal values of the path lengths and task completion time were used for their numeric representation. In this case all partial criteria shall reach maximum, and that is why integral criterion is calculated by summation according to the formula (1.21). Partial criteria averaged values are given in Table 1.

Regarding the partial criteria, the following maximum values were reached:
- safety criterion 0.838 (combined planning method using unstable modes and virtual goal point);
- path length 0.03205 (potential-field method 3);
- task completion time 0.07008 (DVH method in a traditional basis and in formal neuron basis);
- mission success coefficient is 1 (potential-field method 3, control method using fuzzy systems, DVH method in a traditional basis and in formal neuron basis, neural network hybrid method, genetic search with mapping, combined planning method using unstable modes and virtual goal point).

Further, using maximum values of partial criteria, let us perform normalization of these criteria. As a result we receive values of normalized partial criteria and an integral criterion calculated according to the formula (1.21). These results are given in Table 2.

Results of integral criterion calculation are represented in the last column of Table 2. It shall be noted that DVH method in a traditional and formal neuron

Table 1 Partial Criteria Averaged Values

Methods	s_m	P_L	t_m	F
		Criteria		
Potential-field method 1	0.104	0.01975	0.01593	0.6
Potential-field method 2	0.028	0.00665	0.00532	0.2
Potential-field method 3	0.37	0.03205	0.02593	1.0
Potential-field method 4	0.724	0.02595	0.02097	0.8
Potential-field method 5	0.344	0.02	0.01633	0.6
Potential-field method 6	0.576	0.02591	0.02132	0.8
Voronoi diagram method	0.44	0.02052	0.01728	0.6
Reactive navigation method	0.656	0.02412	0.02473	0.8
Control method using fuzzy systems	0.495	0.02507	0.01253	1.0
DVH method in a traditional basis	0.068	0.0313	0.07008	1.0
DVH method in a formal neuron basis	0.068	0.0313	0.07008	1.0
Neural network hybrid method	0.076	0.0317	0.03795	1.0
Genetic search without mapping	0.074	0.01353	0.01103	0.4
Genetic search with mapping	0.565	0.03094	0.02373	1.0
Planning method using unstable modes	0.317	0.02032	0.01964	0.8
Combined planning method using unstable modes and virtual goal point	0.104	0.01975	0.01593	0.6

basis at the set modeling conditions gives similar values of partial criteria, but the difference between them appears during scenes modeling for various initial conditions.

It shall be also noted that good results can be given by the methods based on neural network technologies and combined planning method using unstable modes and virtual goal point. Genetic search is effective when used together with mapping. Quite good results are obtained for neural network hybrid method and control method using fuzzy systems. In case of potential-field methods, there can be seen a substantial scatter of readings which is obviously related to the strong dependence of method features on the selection and parameters of the used attraction and repulsion functions.

It can be noted that if obstacle radius is set with some margin which characterizes safe distance, then safety criterion can be interpreted in a discrete way: if the vehicle came into collision with an obstacle, the safety criterion equals to 0, in other cases it equals to 1. In this regard, mission success and safety criterion duplicate each other to a certain extent. In such task statements integral criterion can be calculated according to the trajectory length, time of task

Table 2 Standardized Values of Partial Criterion and Integral Criterion

Methods	Criteria			
	s_m	p_L	t_m	F
DVH method in a traditional basis	0.0811	0.9766	1.0	1.0
DVH method in a formal neuron basis	0.0811	0.9766	1.0	1.0
Combined planning method using unstable modes and virtual goal point	1.0	0.8038	0.2391	1.0
Genetic search with mapping	0.6742	0.9653	0.3386	1.0
Potential-field method 3	0.4415	1.0	0.3700	1.0
Neural network hybrid method	0.0907	0.9889	0.5415	1.0
Potential-field method 4	0.864	0.8095	0.2992	0.8
Control method using fuzzy systems	0.5904	0.7821	0.1788	1.0
Reactive navigation method	0.7828	0.7526	0.3528	0.8
Potential-field method 6	0.6874	0.8083	0.3043	0.8
Planning method using unstable modes	0.3783	0.6341	0.2804	0.8
Voronoi diagram method	0.5251	0.6402	0.2465	0.6
Potential-field method 5	0.4105	0.6239	0.2330	0.6
Potential-field method 1	0.1241	0.6163	0.2273	0.6
Genetic search without mapping	0.0877	0.4221	0.1573	0.4
Potential-field method 2	0.0334	0.2074	0.0759	0.2

completion, and mission success. Table 3 gives integral performance criteria at the following values of weight coefficients in Eq. (1.21):

$$a_1 = 0; \quad a_2 = 0,3; \quad a_3 = 0,3; \quad a_4 = 0,4.$$

From the last table the conclusion can be made that intelligent technologies based on neural network approaches are highly effective. Also high potential during vehicle path planning tasks is demonstrated by methods based on the usage of unstable modes, fuzzy systems, and genetic search algorithms. Separate attention should be paid to the potential field method 3, which is obviously connected to the successful selection of potential functions.

It should be noted that for solving the application tasks related to safe obstacles avoidance (for example, air defense areas) the best results are shown by combined planning method using unstable modes and a virtual goal point.

The analysis has been carried out for intelligent planning and control systems based on the vehicle's kinematic equations, considering only first model Eq. (1.2). Consideration of vehicle dynamic equations and vehicle's actuators

Table 3 Integral Performance Criterion

Methods	Integral Criterion
DVH method in a traditional basis	0.993
DVH method in a formal neuron basis	0.993
Neural network hybrid method	0.8591
Potential-field method 3	0.8110
Genetic search with mapping	0.7912
Combined planning method using unstable modes and virtual goal point	0.7128
Control method using fuzzy systems	0.6883
Potential-field method 4	0.6526
Reactive navigation method	0.6517
Potential-field method 6	0.6538
Planning method using unstable modes	0.5943
Voronoi diagram method	0.5060
Potential-field method 5	0.4971
Potential-field method 1	0.4931
Genetic search without mapping	0.3339
Potential-field method 2	0.1649

can change the values of partial and integral criteria which require undertaking additional studies. Additionally, further development of the results described in this book is possible for application to three-dimensional dynamic environments. Increase of space dimension requires consideration of additional limitations for the vehicle's functioning modes and makes intelligent planning and control algorithms more complicated. In the dynamically changing environment we have to account for obstacles motion velocities and their direction; vehicle's velocity limitations should also be taken into consideration; the current situation should be forecasted, etc.

Glossary

A*-algorithm A search algorithm of the first best match on the graph, which finds the route with the lowest cost from one vertex (start) to another (target, goal).

Adaptation A control system's capability to change its parameters or structure as a reaction to the changes in plant parameters or external disturbances acting on the plant.

Afferent synthesis A process of comparison, selection, and synthesis of multiple and functionally different afferentations, caused by certain actions onto the objet, and influencing its future behavior.

Artificial potential field Virtual field of potential interaction forces between objects similar to physical fields.

Attractive and repulsive forces Virtual forces generating attraction and repelling of objects in artificial potential filed.

Attractor An attracting manifold in the vehicle's state space.

Autonomy A quality of a system to perform control without external controlling actions.

Axodendritic relation A relation between the neuron-like elements, when information can be passed only from an output of one neural element to an input of another one, unlike the biological neurons, where several closely located inputs or outputs can affect each other.

Behavior A sequence of vehicle's actions formed on the basis of analysis of current situation in environment and vehicle states.

Bifurcation parameter A parameter changing the motion character from stable to unstable.

Bionic neural networking planner A software or hardware tool with an architecture that periodically (1) produces a model representation of the current state of the environment in a form of neural-like structures, and (2) ensures these structures take a subsequent optimal decision about the vehicle's further motion direction.

Cognitron A self-organizing multilayered neural network for solution of image perception and recognition with its architecture, modeling the processes of the real visual images perception (from simple to complicated) in the human visual cortex.

Convolutional neural network, CNN A multilayered artificial feed-forward neural network proposed by Yann LeCun having wide applications in image recognition. It ensures multilevel encoding and recognition of image features with a subsequent probabilistic recognition of the whole image frame.

Crossover Transformation (crossover) of individuals (or their parts) from a parental generation to create individuals of descendants.

Defuzzification A process of producing a quantifiable result in Crisp logic, given fuzzy sets and corresponding membership degrees.

Delaunay Triangulation Triangulation for given set N of points on a plane when all points of N in any triangle, except its vertexes, are located outside circle circumscribed around the triangle.

Distance vector histogram A vector of weighted numeric values proportional to the measured distances to the obstacles that have been detected by a multi-beam locator, where the index of the vector element corresponds to the ordinal number of the active beam.

Dynamic window approach An approach to motion trajectory planning consisting in determination of the vehicle's speed vector admissible in limited time interval with maximization of some objective functions.

Environment uncertainty Situation when there is no information about the parameters and dynamics of changing of environment states.

Expert A specialist able to make (recommend) control decision on the basis of his experience in particular object domain.

External coordinates Independent coordinates specifying vehicle's position and orientation in a certain stationary basic coordinate system.

Fitness function Criteria, on the basis of which comparative estimation of alternative solutions generated by a genetic algorithm is carried out.

Fuzzification Conversion of physical variables into fuzzy inputs, i.e., matching an x variable's physical value and degree $\mu A(x)$ of membership to a fuzzy set A.

Fuzzy controller A controller based on fuzzy logic and implementing fuzzy rules of formation of resultant output variable (scalar or vector) on the basis of analysis of fuzzy input variables or their combination.

Fuzzy inference Obtaining a logical conclusion (control action), corresponding to the current values of input signals with the use of fuzzy knowledge base and fuzzy rules.

Fuzzy logic The logic that operates with fuzzy values and numbers in application of logical operations and is intended for the formalization of expert knowledge in the form of inaccurate or approximate reasoning, which allows more adequate description of uncertain situations.

Fuzzy rule Rule (production) that connects a premise and an effect, which may be fuzzy, with the use of fuzzy logic operations. Creation of fuzzy rules is based on the key knowledge in a particular domain or on expert knowledge.

Fuzzy set A collection of ordered couples composed of elements x from a set X and membership degrees $\mu A(x)$ taking values in the interval [0,1] and indicating the extent (measure) to which the element $x \epsilon X$ belongs to the fuzzy set A.

Fuzzy variable An Element of a linguistic variable determining its basic set, i.e., an interval of values changing.

Gene Indivisible element of an individual.

Generalized work functional Half-defined quadratic performance functional depending on the state variables and optimal control variables.

Generation Individuals Pi of a population generated on the current iteration of an algorithm.

Identification Determining the structure and parameters of a mathematical model ensuring the best match of the model's outputs to those of the plant under similar input actions.

Individual An element Pi of population containing parameters of solution.

Intelligent control system A system performing goal-setting, planning, and motion control under changing conditions without any interaction with an operator or with a higher level system.

Internal coordinates Independent coordinates describing a vehicle's linear and angular velocities in a coordinate system fixed to the vehicle.

Invariance Independence of one or several controlled variables of external disturbances.

Inverse dynamics problem method A system synthesis method based on a search for a control law ensuring equality of higher derivative of the plant's model and higher derivative of the desired differential equation.

Inverse kinematics problem Calculation of internal coordinates using the given external ones.

Lidar A forward-looking multibeam 2D locator (in our case), operating in a horizontal plane, having a limited range and angular resolution specified for each beam.

Linearization Substitution of a nonlinear function for a linear relation at a certain interval.

Linguistic variable A verbal (semantic) expression of some concept in a natural language, the values of which can be determined on fixed sets. The linguistic variable has a term-set, consisting of fuzzy variables.

Local minimum of potential field An area in which a resultant potential force vector is such that a vehicle cannot continue its motion to the goal.

Lyapunov function A positively definite function of plant's state variables used to estimate stability of a motion trajectory.

Lyapunov's direct method A method of stability analysis using sign definite functions.

Mapping A process of creation of a structure describing relative positions of objects in some environments.

Mobile obstacle An obstacle with a changing spatial position.

Mutation Transformation of a parent's individual (or its part) to create an individual of a descendant.

Neocognitron A multilayered self-organizing artificial neural network, which is an evolution of cognitron ideas providing recognition of images invariantly to their location in the image frame.

Neural-like network A mathematical model describing a set of interacting artificial neurons, using the principle of a simplified organization of biological neural networks of living organisms.

Neural-like element (Formal neuron) A simplified formal-logical model of a biological neuron with the values of synaptic coefficients depending on the state of the environment. In the simplest case, its activation function is a threshold function.

Obstacles clusterization Sector partition of vehicle's range of vision. In the considered case, it is assumed to use six clusters (sectors): FL, front left of the vehicle; FR, front right of the vehicle; RU, to the right of the vehicle; RD, rear right of the vehicle; LU, to the left of the vehicle; LD, rear left of the vehicle.

Optimal control A control ensuring the best quality of dynamic processes according to the set functional.

Parents, descendants Elements of populations in genetic algorithms are often referred as parents. Parents are chosen from population on the basis of specified rules and then mixed (crossed over) for descendants (offspring) production.

Population Population P is the set of elements Pi, i = 0,1,2,..,n, where i is the number of generation of a genetic algorithm. Each element Pi of the population, as a rule, is one or several individuals, which are, in fact, alternative ordered and non-ordered solutions.

Population initialization Creation of an initial population (generation of specified quantity of individuals according to a certain algorithm).

Position-path control A control method based on the solution of the inverse dynamics problem in external coordinates that are set in analytical form.

Quadratic form A function on a vector space defined by a homogeneous polynomial of degree two in vector's coordinates.

Rapidly exploring random tree A graph data structure with random exploration elements that is designed for a broad class of path planning problems.

Reference model adaptive system Adaptive control systems using comparison to a reference model or identificational approaches giving the possibility to determine performance function related parameters or characteristics of the control process.

Repeller A repelling manifold in the state space of a vehicle.

Road map A network of 1D curves capturing the connectivity of the vehicle's free space.

Rule base (knowledge base) Set of fuzzy rules formed by an expert allowing the implementation of logical inference in a particular fuzzy situation.

Selection A process through which individuals (alternative solutions) having higher value of objective function (with "better" features) get better opportunity for reproduction of descendants than worse individuals.

Solid body A mechanical system having only translational and rotational degrees of freedom.

Stationary obstacle A resting obstacle.

Stop criterion Value of individuals' fitness function at achievement of which an algorithm stops.

Uncertainty of environment Absence of a priory information about the obstacles present in the environment.

Virtual goal A temporal goal point of vehicle, during motion to which a real goal is ignored.

Voronoi diagram Partitioning of a plane with N points into a set of convex polygons in such a way that each of them contains seed, and any point inside a polygon is closer to its seed than to any other.

Index

Printed in the United States
By Bookmasters